24th Condensed Matter Days National Conference (CMDAYS 2016)

Journal of Physics: Conference Series Volume 765

Mizoram, India
29 – 31 August 2016

Editor:

R.K. Thapa

ISBN: 978-1-5108-3182-7
ISSN: 1742-6588

Printed from e-media with permission by:

Curran Associates, Inc.
57 Morehouse Lane
Red Hook, NY 12571

CURRAN ASSOCIATES INC. proceedings.com

Some format issues inherent in the e-media version may also appear in this print version.

Printed by Curran Associates, Inc. (2016)

For permission requests, please contact the Institute of Physics
at the address below.

Institute of Physics
Dirac House, Temple Back
Bristol BS1 6BE UK

Phone: 44 1 17 929 7481
Fax: 44 1 17 920 0979

techtracking@iop.org

Additional copies of this publication are available from:

Curran Associates, Inc.
57 Morehouse Lane
Red Hook, NY 12571 USA
Phone: 845-758-0400
Fax: 845-758-2633
Email: curran@proceedings.com
Web: www.proceedings.com

Table of contents

Volume 765

24th Condensed Matter Days National Conference (CMDAYS2016)
29–31 August 2016, Mizoram, India

Accepted papers received: 5 October 2016
Published online: 26 October 2016
Preface

Papers

Journal of Physics: Conference Series **765** (2016) 011001 doi:10.1088/1742-6596/765/1/011001

Introduction

The 24th edition of Condensed Matter Days (CMDAYS) 2016, a National Conference had been decided to be held at Physics Department, Mizoram University, Aizwal, Mizoram, India during 29-31 August 2016. This decision was taken by the General Body meeting of the CMDAYS on 28 August 2015 at Viswa Bharati, Shanti Niketan, West Bengal, India and Prof. R.K. Thapa was proposed as the Convener of CMDAYS-2016. Initiated by the Institute of Physics, Bhubaneswar, Odisa. The CMDAYS conference is held annually in the last week of August. The main objective of the conference was to bring all the researchers/scientists working in the field of Condensed Matter Physics, or related topics, together on a single platform. In this way, they can present, share and discuss their research findings and further plan collaborative works in future. The conference topics were on the theory and experimental research works done on Strongly correlated systems, Soft condensed matter, Magnetism and Magnetic materials, Disordered systems, Phase transition, Materials for energy harvesting, Nanomaterials and applications, Dielectrics and Ferroelectrics, Optoelectronics and devices, Semiconductors and devices, Biophysics, Biomaterials and composites, Superconductivity, Thin films and devices. It was open to all researchers from the research institutes, universities and colleges.

Until the last date 1st June 2016, we have received 1 plenary lecture, 3 Keynote lectures, 8 invited talks and 55 oral contributed papers. In total, there were 10 technical sessions to complete all the contributed papers along with the invited talks. Sessions were very interesting with the young participants interacting extensively with the senior scientists and everybody enjoyed the conference period with two cultural programmes.

On the last day after the closing function, a local tour programme was arranged for all the outside participants.

We are grateful to Prof. R. Lalthantluanga, Vice Cahncellor, Mizoram University, Aizawl, Mizoram for extending financial support to hold this conference. We are also thankful to North Eastern Council, (NEC) Shillong, Science and Engineering Research Board (SERB-DST) (Delhi) and Defence Research and Development Organisation (DRDO, Delhi) for financial support without which this conference would not have been possible.

30th September 2916

(Prof. R.K.THAPA)

Convener, CMDAYS-2016
&
Chief Editor
IOP JPCS

Journal of Physics: Conference Series **765** (2016) 011001 doi:10.1088/1742-6596/765/1/011001

NATIONAL ORGANISING COMMITTEE

Dr. Arunava Chakrabarty, Professor, Kalyani University
Dr. Asok K. Sen, Retd. Professor, SINP, Kolkata
Dr. Bikash C. Gupta, Assoc. Professor, Viswa-Bharati
Dr. Bikash K. Chakrabarti, Professor, SINP, Kolkata
Dr. D. Mohanta, Assoc. Professor, Tezpur University
Dr. Deepali Sarkar, Professor, Gauhati University
Dr. K. Kundu Retd. Professor, IOP, Bhubaneswar
Dr. N. C. Mishra, Professor, Utkal University
Dr. S. K. Rout, Assoc. Professor, BIT, Mesra
Dr. S. M. Bhattacharjee, IOP, Bhubaneswar (Chairman, CMDAYS)
Dr. S. Panigrahi, Professor, NIT, Rourkela
Dr. Sudipta Bhattacharya, Assoc. Professor, CU
U. De, Retd. Professor, VECC, Kolkata

LOCAL ORGANISING COMMITTEE

Prof. R. K. Thapa, Physics, MZU, Convener, CMDAYS-2016
Prof. R. C. Tiwari, Physics, MZU
Prof. Zaithanzauva Pachuau, Physics, MZU
Prof. S. Rai, Physics, MZU
Prof. Diwakar Tiwari, Physics, MZU
Prof. Jamal Hussain, Mathematics, MZU
Dr. B. Muthukumaran, Chemistry, MZU
Dr. Hranghmingthanga, Physics, MZU
Dr. B. Lalremruata, Physics, MZU, Secretary, CMDAYS-2016
Dr. Lalthakimi Zadeng, Physics, MZU, Treasurer, CMDAYS-2016

ABSTRACT COMMITTEE

Prof. R. K. Thapa
Dr. B. Lalremruata
Dr. Sandeep

Dr. Dibya Prakash Rai
Himanshu Joshi

SUB-COMMITTEES

Registration Committee
Rebecca Lalngahawmi
J. Lalnunthari
Andrew Lungtiawia
Sanjay Singh
Lalhriatpuia Hnamte

Accomodation and Catering Committee
Himanshu Joshi
Rebecca Lalngaihawmi
Sanjay Singh
J. Lalnunthari

Transport Committee
Lalrinmawia Punte
Lalhriatpuia Hnamte
T. Malsawmtluanga
K. Neupane

Printing & Poster Committee
H. Lalremruata
Rebecca Lalnungtluangi
Sanjay Singh
Lalhriatpuia Hnamte

Sound System & Hall arrangements Committee
V. Malsawma
Andrew Langtiawia
Rebecca Lalnungtluangi
J. Lalnunthari
T. Malsawmtlaunga
K. Neupane

Publicity Committee
T. Mala
Dr. Sandeep

Journal of Physics: Conference Series **765** (2016) 011001 doi:10.1088/1742-6596/765/1/011001

OUR KIND SPONSORERS ARE ...

Mizoram University

Science & Engineering Research Board
Department of Science & Technology
Government of India

SERB (DST)

North Eastern Council
(NEC)
(H.Q. Shillong)

NEC

DRDO

... THANKS TO THEM !!!

Journal of Physics: Conference Series **765** (2016) 011001 doi:10.1088/1742-6596/765/1/011001

Editorial Board

- **Dr. B. Lalremruata**
  Physics Department
  Mizoram University
  Aizawl 796 004
  Mizoram India
  Email: marema_physics@mzu.edu.in

Executive Editors

- **Dr. Dibya Prakash Rai**
  Physics Department
  Pachhunga University College
  Mizoram University
  Aizawl 796001 Mizoram India
  Email: dibyaprakashrai@gmail.com

- **Dr. Sandeep**
  Condensed Matter Theory Research Group
  Physics Department
  Mizoram University
  Aizawl 796 004 Mizoram India
  Email: sndp.chettri@gmail.com

Associate Editors

- **Prof. Leszek Jurczyszyn**
  Institute of Experimental Physics
  University of Wroclaw
  Plac Maxa Borna 9 Wroclaw, Poland
  Email : lju@alpha.ifd.uni.wroc.pl

- **Prof. T. P. Sinha**
  Physics Department
  Bose Institutte
  93/1, Acharya Prafulla Chandra Road
  Kolkata 700009 India
  Email : sinha_tp@yahoo.com

- **Prof. S. Dorendrajit Singh**
  Physics Department
  Manipur University
  Imphal, Manipur India
  Email : dorendrajit@yahoo.co.in

- **Dr. V. Madhurima**
  Physics Department
  Central University of Tamil Nadu
  Thiruvarur Tamil Nadu
  Email : madhurima.v@gmail.com

- **Prof. Deepali Sarkar**
  Physics Department
  Gauhati University
  Guwahati Assam India
  Email:deepalisarkar@gmail.com

- **Prof. Pradip Kumar Mandal**

  Physics Department
  North Bengal University
  Raja Rammohunpur
  Siliguri, Darjeeling
  Email : mandal_pradip@yahoo.com

- **Dr. Dambarudhar Mohanta**
  Physics Department
  Tezpur University
  Tezpur, Assam, India
  Email : best@tezu.ernet.in

- **Dr. B. Indrajit Sharma**
  Physics Department
  Assam University
  Silchar Assam India
  Email : indraofficial@rediffmail.com

- **Dr. Javad Hashemifar**
  Physics Department
  Quantum Simulation of Materials Group
  Isfahan University of Technology
  Isfahan, Iran
  Email : hashemifar@cc.iut.ac.ir

- **Dr. M. P. Ghimire**
  IFW-Dresden, Helmholtzstraße 20, D-01069
  Dresden, Germany
  E-mail: ghimire.mpg@gmail.com

Journal of Physics: Conference Series **765** (2016) 011001 doi:10.1088/1742-6596/765/1/011001

- **Dr. Amit Shankar**
  Physics Department
  North Bengal University
  Raja Rammohunpur
  Siliguri, Darjeeling
  Email : amitshan2009@gmail.com

- **Dr. (Ms) Lalthakimi Zadeng**
  Physics Department
  Mizoram University
  Aizawl 796 004 Mizoram India
  Email: kimizadeng@gmail.com

Members :

- Dr. Dhanoj Gupta, NFRI, Gunsan, South Korea.

- Dr. Dhiraj Brahman, Department of Chemistry, St. Joseph's College, Darjeeling 734104 India

- Dr. Hrangmingthanga
  Physics Department
  Mizoram University
  Aizawl 796004
  Mizoram India

Journal of Physics: Conference Series **765** (2016) 011001 doi:10.1088/1742-6596/765/1/011001

Condensed Matter Theory Research Group, Department of Physics, Mizoram University

The researches in condensed matter physics/solid state physics under the leadership of Prof. R. K. Thapa has made considerable progress in Mizoram (India) for the last 30 years. With his continuing efforts, he has created facilities for researches in Theoretical Condensed Matter Physics (Solid State Physics) through various research projects from agencies like U.G.C., C.S.I.R, D.A.E and D.S.T. These facilities had been now availed by the students and researchers/teacher fellows, as well scientists from other parts of the country. The name of the research group is CONDENSED MATTER THEORY RESEARCH GROUP (CMT Research Group).

The aims and objectives of the CMT Research Group is to do research investigations in the field of Solid State Physics and allied areas of physics that bears a relation to material science and its properties. The members of the group is also involved in imparting knowledge in computational physics. Presently there are 15 members.

The Main Research Fields of CMT group

- Photoemision and Photofield Emission studies from metals, semiconductors and superconductors.
- Studies of energy band structure, elastic, thermoelectric, optical and magnetic properties of Transition metal nitrides, carbides, skutterdites, rare earth and Heusler compounds.
- Study of single and double perovskite

Few Achievements by members

- No. of Ph.Ds from the group = 13
- No. of research papers published in International journals >150
- Reviewers of Elsevier, IOP, RSC (London), IJP(India), World Scientific etc. journals.
- Research projects completed = 10
- Post octoral studies abroad = 4
- Alexander von Humboldt fellow = 1
- ICTP fellow to Isfahan University of Technology, Iran.
- Attended Workshop and Conference by Ph.D. student in Isfahan University of Technology, Iran (2016).
- JNCASR Visiting Fellow to NPL, Delhi.
- INSA Visiting Fellow to UK (1998), Germany (2011) and Poland (2016)

Journal of Physics: Conference Series **765** (2016) 011001 doi:10.1088/1742-6596/765/1/011001

Photo-1: Photo of the Participants in CMDAYS-2016 with Prof. R. Lalthantluanga, Vice Chancellor, Mizoram University, Aizawl, Mizoram, India.

Photo-2: Photograph of Condensed Matter Theory Research Group with Prof. N.Kar (Grand Supervisor).

Names (Left to Right): Himanshu, Malsawmtluanga, Nkhimananda,Nirmala Devi, Prof. N.Kar, Prof. R.K.Thapa, Rebecca, Dr. Sandeep, Lalhriatpuia and Dr. Amit

Journal of Physics: Conference Series **765** (2016) 011001 doi:10.1088/1742-6596/765/1/011001

Photo-3: Photograph of Condensed Matter Theory Research Group with Prof. R.K.Thapa (Supervisor & Head)).

Names (Left to Right): Malsawmtluanga, Dr. Amit, Dr. D.P. Rai, Nirmala Devi, khimananda, Dr. Sandeep, Prof. R.K.Thapa, Himanshu, Lalhriatpuia and Rebecca.

Peer review statement

All papers published in this volume of *Journal of Physics: Conference Series* have been peer reviewed through processes administered by the proceedings Editors. Reviews were conducted by expert referees to the professional and scientific standards expected of a proceedings journal published by IOP Publishing.

De-pinning of contact line of droplets on rough surfaces

V Madhurima[1] and K Nilavarasi[2]

[1,2]Department of Physics, School of Basic and Applied Sciences, Central University of Tamil Nadu, Thiruvarur – 610101, India

*madhurima@cutn.ac.in; #nilavarasikv@gmail.com;

Abstract. The present study reports the formation of self-assembled droplet pattern on the PDMS polymer coated over grooved side of DVD under saturated vapours of alcohols. Comparison of the results with breath figures formed over unconstrained side of DVD is made. Four different environments namely methanol, ethanol, 2-propanol and n-butanol are used for the analysis. It is observed that the pattern formation occurs with methanol and ethanol vapours and not with 2-propanol and n-butanol. The difference is pattern formation with different alcohols is attributed to the variation in chain length and the presence of hydrophobic groups in alcohols, as given by Traube's rule. The distortion of patterns over constrained surface is attributed to the depinning of contact lines.

1. Introduction

The study of dynamics of a droplet's contact line on structured surfaces is of importance for its applications in self-cleaning [1], anti-corrosion [2], bio-sensing [2] and so on. The contact line dynamics is greatly influenced by the presence of surface heterogeneities [3]. These heterogeneities acts as droplet pinning sites. Although the dynamics of the contact line for hydrophobic/super-hydrophobic surfaces were extensively studied, the contact line dynamics during the phase transition process like evaporation and condensation were relatively less studied [2, 4].

Breath figure is one such process involving the phase change process of evaporation and condensation to form self-assembled droplet patterns [6, 7, 8, 9, 10, and 11]. Usually breath figures are formed with super saturated vapor of water. Ding et al., was the first to report the formation of self-assembled droplet pattern using non-aqueous vapors of ethanol and methanol [12]. They proved that the breath figure method is not only suitable for water vapor but is also compatible with organic solvents like ethanol and methanol. But the influence of constraints on the underlying surface on the formation of breath figures using non-aqueous vapor is so far not reported to the best of our knowledge.

Alcohols are the simplest form of amphiphilic molecules having hydrophobic and hydrophilic groups [13]. These hydrophobic and hydrophilic group of the alcohols play an important role in deciding the contact area of the droplets. When a droplet of liquid is impinged on the surface (preferentially solid/soft solid surface), the hydrophobic group of an added alcohol preferentially goes to the surface [13]. The hydrophobicity of the alcohols increases with increase in the chain length of alcohols and at the interface, surface tension decreases with increase in number of CH_2 group according to Traube's rule [14].

The present paper deals with the experimental verification of the effect of hydrophobic interactions in the formation of self-assembled droplet formation and Traube's rule of surface tension at interfaces. The effect of contact line dynamics of these alcohols on the constrained surface is also investigated.

2. Theory

When a liquid droplet is placed on a solid surface having heterogeneities, the contact line can stay pinned until the contact angle reaches a critical angle [17]. The contact angle for smooth surface under thermodynamic equilibrium conditions is defined by Young's equation, which is given as [15],

$$cos\theta_y = \frac{\gamma_{sv}-\gamma_{sl}}{\gamma_{lv}} \tag{1}$$

where γ_{sv} , γ_{sl} and γ_{lv} are the solid-vapour, solid-liquid and liquid-vapor interfacial tensions respectively and θ_y is the Young's contact angle which is the intrinsic contact angle of the surface at the three phase contact line. For a heterogeneous rough surface, the contact angle can vary over a finite range [16, 17]. On a smooth surface, the contact line will not advance (recede) until the contact angle reaches a critical value of the advancing (receding) contact angle. The difference between the advancing and receding contact angle is called contact angle hysteresis, which characterizes the strength of pinning of contact line at the asperities of the rough surface [16].

After pinning the droplets start to evaporate. There are two different modes of evaporation of a liquid droplet on a polymer surface, namely, the constant contact radius (CCR) mode and the constant contact angle (CCA). The CCR mode is characterized by the pinned contact line and decreased contact angle whereas the CCA mode is characterized by the receding contact line and constant contact angle [2].

The behavior of droplet evaporation is also affected by the surface morphology and chemical composition. Orejon et al.[18] in their investigation of the dynamics of the three-phase contact line of evaporating droplets on different surfaces with varying substrate hydrophobicity proposed that the depinning of contact line on rough surfaces occurs when the unbalanced Young's force is large enough to overcome an intrinsic energy barrier arising from surface roughness. CCR-CCA transition, a consequence of the pinning-depinning phenomenon was observed during the evaporation on the surfaces with relatively large stripe widths [19].

3. Experimental details

3.1. Materials and methods

Commercially available digital versatile disk was used as substrate in the present study. In the present study, poly-dimethyl siloxane (PDMS) from Sigma-Aldrich and chloroform (purity 99.9%) purchased from Merck (Emplura) were used as polymer and solvent respectively. Methanol, ethanol, n-butanol and 2-propanol purchased from Merck (Emplura) were used as environment.

The aluminium layer in the DVD were peeled off using forceps and the side facing the aluminium layer was used as constrained surface consisting of parallel grooves. The other side was used as the smooth surface.

For the preparation of patterned films using the supersaturated vapor of methanol, 5 ml of methanol is added to the petri-dish and kept inside the glass chamber. The polymer solution was prepared and casted on the surface of the disk. The casted disk was then placed inside the chamber containing methanol vapors. After complete evaporation of solvent, the films were characterized using confocal laser scanning microscope (CLSM). Same procedure is followed to prepare patterned films with other vapors namely ethanol, 2-propanol and n-butanol.

3.2. Mechanism of formation of self-assembled droplet patterns

In the beginning of the formation of self-assembled droplet pattern, the non-aqueous vapor starts to condense due to evaporation of the solvent. The droplets condense on the polymer solution surface.

Spreading and dissolving of droplets in the polymer solution shouldn't take place if pores are to be formed. The spreading behavior of the droplet on the surface is defined using spreading co-efficient as,

$$S = \gamma_s - (\gamma_l + \gamma_{ls}) \tag{2}$$

where γ_s and γ_l are the surface tension of the solution and condensed liquid respectively. γ_{ls} is the interfacial tension of the liquid and solution. S > 0 indicates spreading behavior is more and S < 0 indicates non-spreading behavior of the liquid. Once the solvent and liquid completely evaporates, the polymer film is left with array of pores.

4. Results and Discussion

The patterned films on smooth and constrained surfaces using methanol, ethanol, 2-propanol and n-butanol vapours were characterized for surface morphology using CLSM and are shown in figure 1. With methanol and ethanol vapours, the films were patterned with pores. The smooth surface showed a hexagonal array of self-assembled droplet pattern as expected due to Marangoni convection and formation of a lubricating thin polymer layer as a result of instant precipitation of the polymer [21]. The instant precipitation prevents the droplets from coalescing with each other thus forming hexagonal pore

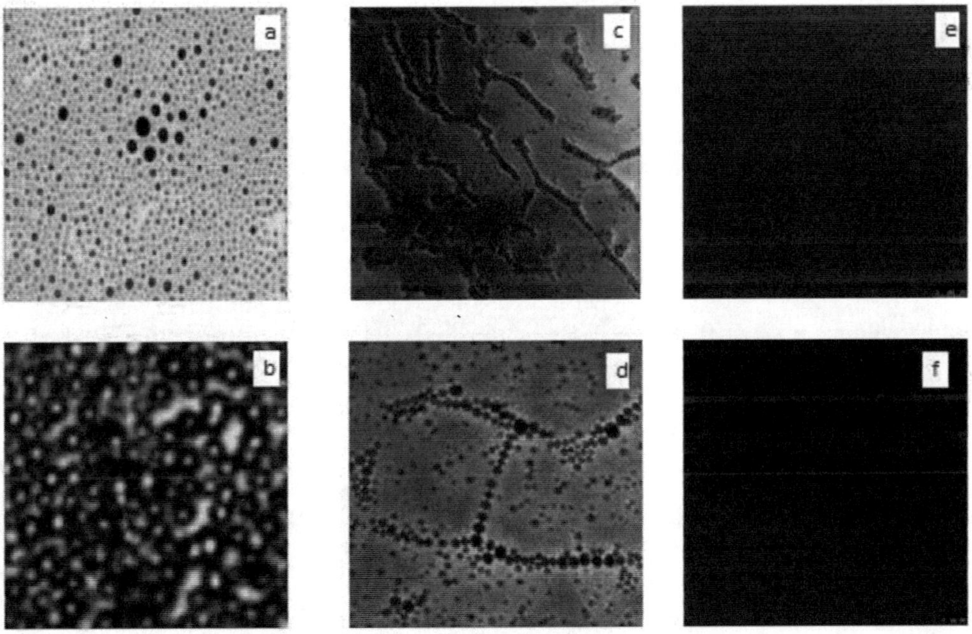

Figure 1. Self-assembled droplet patterns formed on a. smooth surface with ethanol vapor, b. smooth surface with methanol vapor, c. constrained surface with ethanol vapor d. constrained surface with methanol vapor, e. surfaces with propanol vapor and f. surfaces with butanol vapor.

patterns. The CLSM images of grooved surface showed a different pattern of pores. This distortion in pattern formation is ascribed to the depinning mechanism of contact lines. This depinning occurs due to the presence of constraints on the underlying surface [20, 3].

For a droplet on a rough surface having number of grooves parallel to each other, the contact line can align in two ways. It can lie either parallel to the grooves and get pinned or lies at an angle to the grooves. In the latter case, the contact line is displaced continuously without any pinning [3]. This prevents the self-assembly of droplets into a hexagonal pattern. As a result, a distorted hexagonal ring of pores with cluster of pores inside is observed.

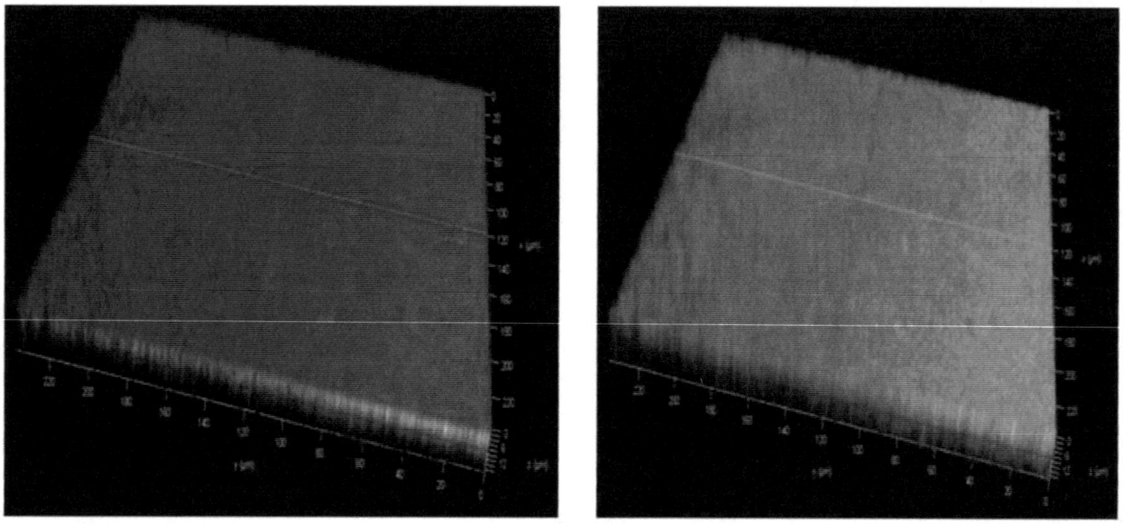

Figure. 2. 3D images of surfaces with propanol and n-butanol vapours. Confirming the presence of polymer film over the substrate.

The CLSM images of surfaces in 2-propanol and n-butanol environment doesn't show any patterns and is shown in figure 1. The 3D images of the polymer film formed over the substrate is shown in figure2. For 2-propanol and n-butanol, there is an increase in CH_2 group. This increase in CH_2 group reduces the surface tension which in turn affects the contact area of the droplet. The increase in droplet contact area favours the coalescence of droplets. On complete evaporation of the solvent, a layer of polymer film without any significant patterns is observed. This result also proves the Traube's rule relating surface tension of organic compounds to the number of hydrocarbon CH_2 groups present in the molecule. According to Traube's rule, for every extra CH_2 group, the reduction in surface tension triples [14].

5. CONCLUSION

The formation of a self-assembled droplet pattern on the PDMS polymer coated over grooved side of DVD under saturated vapors of alcohols is reported. It is observed that with the increase in alkyl chain length of alcohols, the formation of droplet patterns decrease. Hexagonal pattern of self-assembled droplets on smooth surface was observed for surfaces subjected to methanol and ethanol vapors. A distorted hexagonal ring with cluster of pores inside was observed for grooved surface under same vapor conditions. The distortion in pattern formation is attributed to the depinning effect of contact line on rough surface. With 2-proppanol and n-butanol, no patterns of droplets were observed. This confirms Traube's rule of decrease in surface tension with increase in number of CH_2 groups.

References

[1] Nosonovsky M and Bhushan B 2007 Nano Lett.**7** 2633
[2] Chen X, Ma R, Li J, Hao C, Guo W, Luk B L, Li S C, Yao S and Wang Z 2012 PRL **109** 116101
[3] de Gennes Wetting P G 1985 statics and dynamics Reviews of Modern Physics **57(3)**, 827-863
[4] Gelderblom H, Marin A G, Nair H, Van Houselt A and Lefferts L 2011 Phys. Rev. E **83**, 026306
[5] Zheng Y, Kubowaki Y, Kashiwagi M and Miyazaki K 2011 J. Mechanical Science

and Technology **25 (1)** 33
[6] Bunz U H F 2006 Adv. Mater. **18** 973 – 989
[7] Stenzel M 2002 Aust. J. Chem. **55** 239 – 243
[8] H. Ma, J. Hao, Chem. Soc. Rev., 40, 5457 – 5471, (2011)
[9] Pierre Escale, Laurent Rubatat, Laurent Billon, Maud Save, European Polymer Journal 48, 1001 -1025,(2012)
[10] A. Steyer, P. Guenoun, D. Beysens, C. Knobler, Phys. Rev. B, 42, 1086 – 1089, (1990)
[11] O. Pitois, B. Francois, Colloid Polym. Sci. , 277, 574 – 578, (1999)
[12] Jianun Ding, Aijuan Zhang, Hua Bai, Lei Li, Jian Li and Zhi Ma, Soft Matter, 9, 506-514, (2013)
[13] Deepti Ballal and Walter G. Chapman, J. Chem. Phys.139, 114706 (2013)
[14] Ruth H Aranow and Louis Witten, J. Chem. Phys. 28, 405 (1958)
[15] T. Young, Phil. Trans. Roy. Soc. (London) 95, 65–87 (1805).
[16] Yevgeniy V. Kalinin, Viatcheslav Berejnovand Robert E. Thorne, Langmuir. 25(9): 5391–5397 (2009)
[17] A. Marmur, Adv. Colloid Interface Sci. 19, 75–102 (1983).
[18] Orejon, D. Sefiane, K.; Shanahan, M. E. R. Langmuir, 27, 12834-12843 (2011).
[19] Wang, F.C, Wu, H.A. Soft Matter,9, 5703-5709 (2013).
[20] K Nilavarasi and V Madhurima, arXiv:1605.02942 [cond-mat.soft], (2016)
[21] Mohan Srinivasarao, David Collings, Alan Philips, Sanjay Patel. Science, 292:79-82, (2001).

Journal of Physics: Conference Series **765** (2016) 012002 doi:10.1088/1742-6596/765/1/012002

Angle resolved photoemission spectroscopy and surface states

Nikhiles Kar

Physics Department, RKM Vivekananda University, Kolkata, West Bengal, India

*Email: amitshan2009@gmail.com

Abstract. Angle Resolved Photo Emission Spectroscopy (ARPES) has been a very effective tool to study the electronic states of solids, from simple metals to complex systems like cuprate superconductors. For photon energy in the range of 10 - 100 eV, it is a surface sensitive process as the free path of the photo emitted electrons is of the order of a few lattice parameters. However to interpret the experimental data one needs to have a theoretical foundation for the photoemission process. From the theory of photoemission it may be seen that one can get information about the state from which the electron has been excited. As the translational periodicity is broken normal to the surface, a new type of electron state in the forbidden energy gap can exist localized in the surface region. ARPES can reveal the existence and the property of such surface states. We shall also discuss briefly how the electromagnetic field of the photons are influenced by the presence of the surface and how one can try to take that into account in photoemission theory.

1. Introduction

Angle Resolved Photoemission Spectroscopy (ARPES) has been an important technique for studying the electronic states of metals, semiconductors (both clean and adsorbate covered) as well as more complex systems e.g. High Temperature cuprate superconductors. The development in making clean surfaces, working in ultra-high vacuum environment and the avail- ability of photons with continuously varying wave lengths from synchrotron sources led to accurate measurements. The schematic experimental setup is shown in the figure below.

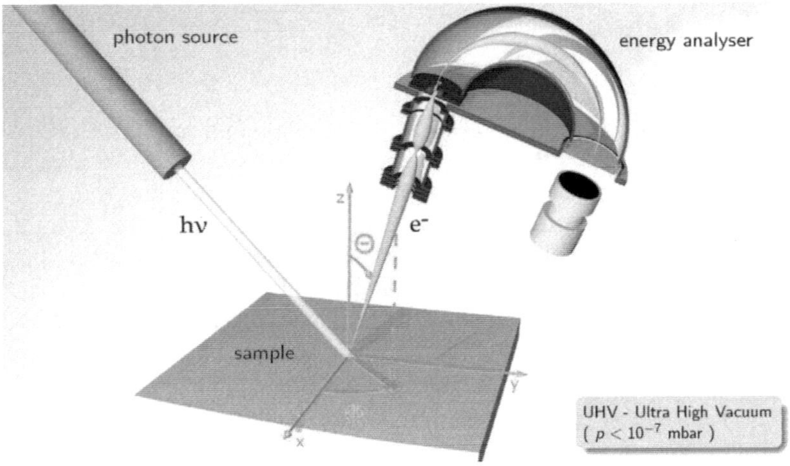

Figure 1: Schematic Experimental Setup

There are a large number of variables in the experiment e.g.
(i) Electron energy
(ii) Direction of emission (angle at which the electron is collected) (iii) angle of incidence of the photons.
(iv) Photon energy
(v) Polarization of the photon beam

1.1. *Photoemission Theory*

The photocurrent in the independent electron approximation by application of Fermi's Golden Rule

$$\frac{dj(E)}{d\Omega}=\frac{2\pi}{\hbar}\sum|\langle\psi_f|H'|\psi_i\rangle|^2\,\delta(E-E_f)\delta(E_f-E_i-\hbar\omega)f_0(E_i)[1-f_0(E_f)] \qquad (1)$$

$|\psi_i\rangle and|\psi_f\rangle$ are the initial and final states and $H' = \frac{e}{2mc}(\boldsymbol{A.p} + \boldsymbol{p.A})$ is the perturbing Hamiltonian up to first order in the photon field vector potential **A**. The two Fermi distribution functions f_0 ensure that the initial state is occupied while the final state is unoccupied. The final state energy E_f is greater than the initial state energy E_i by $\hbar\omega$ where ω is the photon frequency.

$|\psi_f\rangle and|\psi_i\rangle$ in equation (1) are in principle many body states with N and (N-1) electrons (N is the total number of electrons). Since it is virtually impossible to deal with a system of ~ 10^{23} particles, $|\psi_f\rangle and|\psi_i\rangle$ are almost always taken as single particle states in one electron approximation.

The simplest approximation is to consider that as the electron coming to the detector may be taken to be a free electron, there is no structure from the final state; the vector potential is also taken to be constant and so all the features seen in the photocurrent are from the initial state density of states. However such an approach is too simplistic and one needs to have a better interpretation. One way of doing that is the so called three step process.

In the first step one calculates the matrix element $\langle \psi_f | H' | \psi_i \rangle$ between the initial band state $\langle \psi_i |$ and the final band state $\langle \psi_f |$ - both calculated by some band structure method, e.g. self-consistent LAPW or LKKR method in density functional formalism. The second step is to consider the travel of the electron to the surface and the third step is the transmission through the surface. This is discussed in more detail in references [1] and [2].

In this review. we shall be focusing our attention to the so called one- step calculation, which starts by recognising that the presence of the surface destroys the three dimensional periodicity intrinsic to band structure calculations. The wave-vector k in Bloch's theorem $\psi(\mathbf{r+R}) = \exp(i\mathbf{k.R})\psi(r)$ is a vector with three real components only because of three-dimensional period icity. In the presence of the surface we may have two-dimensional periodicity and while the components of **k** parallel to the surface, $\mathbf{k}_{||}$ would be real - the component $\mathbf{k}_{\perp}$, perpendicular to the surface need not be real, but would in general be complex. In the following we discuss the photocurrent calculation in such a context.

1.2. Final State

For the final state, we have an electron (free electron) going to the detector. Constructing the free electron wave function by starting with a band wave-function and then matching at the surface is a difficult problem. So the usual technique is to take a different approach.

Let us consider an electron coming from the detector $(\psi \sim e^{(ik_f.r)})$. After it is incident at the surface there will be reflected and transmitted waves with proper wave function matching. This is the Low Energy Electron Diffraction (LEED) problem and for that problem the theory and computational techniques are easily available (see for example [5]). Now, if we consider that state to be time-reversed, we get a free electron going to the detector! The 'Time Reversed LEED state' discussed by Liebsch [6], Pendry [7] and others (see [3]) has been the standard technique for final state calculations. Pendry has shown that one can write down the photocurrent in terms of an appropriate Green's function - the multiple scattering and lifetime effects can then be included in the calculation. Using atomic units Pendry derived an expression for the current as

$$I(k_{||}, E_f) = K_{2\perp}\rho(k_{||}, E_i)|\langle k_{||}|G_2^+ H'|\psi_f\rangle|^2 \quad (2)$$

$$\langle r | k_{||} \rangle = \frac{1}{L}\delta(z - Z)\exp(ik_{||}.r_{||}) \quad (3)$$

Where Z is the plane of observation and L is the dimension of the box containing the system. Pendry [3] showed that this is equivalent to the " Time Reversed LEED State' formalism and can be incorporated in the photoemission calculation.

1.3. Initial State

The initial state is usually that of an electron in a band. However, as mentioned before, because of the surface, the three dimensional periodicity is not valid. Assuming a semi-infinite surface would allow us to have two dimensional periodicity. This implies that the wave-vector k will have its component $k_{||}$ as real but the normal component $k_{\perp}$ may now be complex! Normally one may think of computing the initial state wave function in a density functional framework but the standard methods based on three dimensional periodicity will not be appropriate. One has to do a layer by layer calculation (with the layers parallel to the surface plane) to be consistent with the final state calculation.

The wavefunction for a complex band with band index α may be written as, (z taken as the direction perpendicular to the surface)

$$\phi_\alpha = \sum_g \phi_{g\alpha}(z)\exp i(k_{\|}.r_{\|}) \quad \text{II} \tag{4}$$

where, g is a set of 2 dimensional reciprocal lattice vectors. In the region outside the surface

$u_g(z)$ is a function decaying exponentially away from the surface plane (z=0).
The wavefuction when properly matched across the surface plane will give the initial state wave function.

1.4. Perturbing Hamiltonian

The perturbing Hamiltonian $H' = \frac{e}{2mc}(\boldsymbol{A}.\boldsymbol{p} + \boldsymbol{p}.\boldsymbol{A})$ can be simplified by using the commutator relation $[\mathbf{p}, \mathbf{A}] = -i\hbar\nabla.A$ and choosing the gauge $\nabla.A = 0$, one may write

$$H' = -(e/mc)A.p \tag{6}$$

Usually the spatial variation of A is neglected and the vector potential A is taken as a constant. We shall discuss this point in more detail later.

1.5. Example of Photocurrent Calculation

Photoemission computer programs were developed to include the matrix element calculation taking care of the various terms as discussed above. The details of computational technique of the photocurrent calculation have been given by Hopkinson, Pendry and Titterington [8]. Kar [9] used the program to calculate the photoemission spectra for bromine adsorbed on copper. The experimental data for the spectra were from Richardson and Sass [10]. We show below the experimental data for normal photoemission from clean Cu (100) surface and with a c(2×2) Bromine over layer. B, C and D identify the new features due to the adsorption of bromine. The theoretically calculated spectra for normal emission is given in the next figure (from [9]). The next two figures show the experimental and calculated data for angles of emission from 0 to 50 degrees at 10 degree interval. It may be seen that the theoretically calculated spectra qualitatively reproduce the new features for normal photoemission and also the trends when the angles of detection are changed,

Journal of Physics: Conference Series **765** (2016) 012002 doi:10.1088/1742-6596/765/1/012002

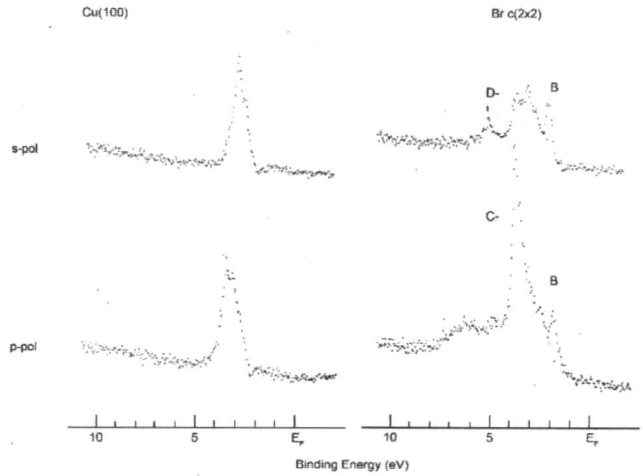

Figure 2: Experimental data for normal photoemission: s and p polarisation

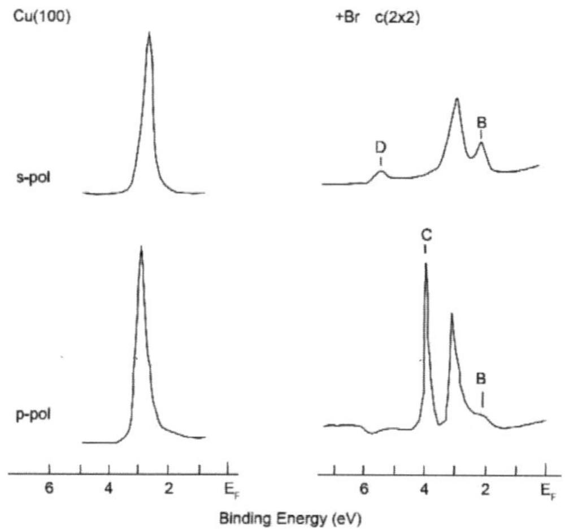

Figure 3: Calculated spectra for normal photoemission: s and p polarization

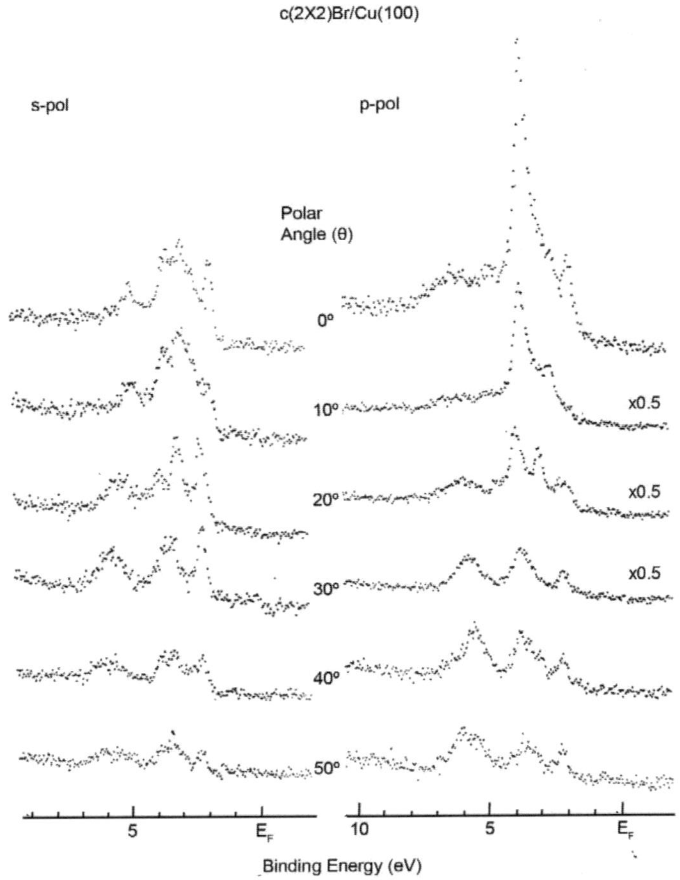

Figure 4: Experimental data 0 to 50 degrees: s and p polarization

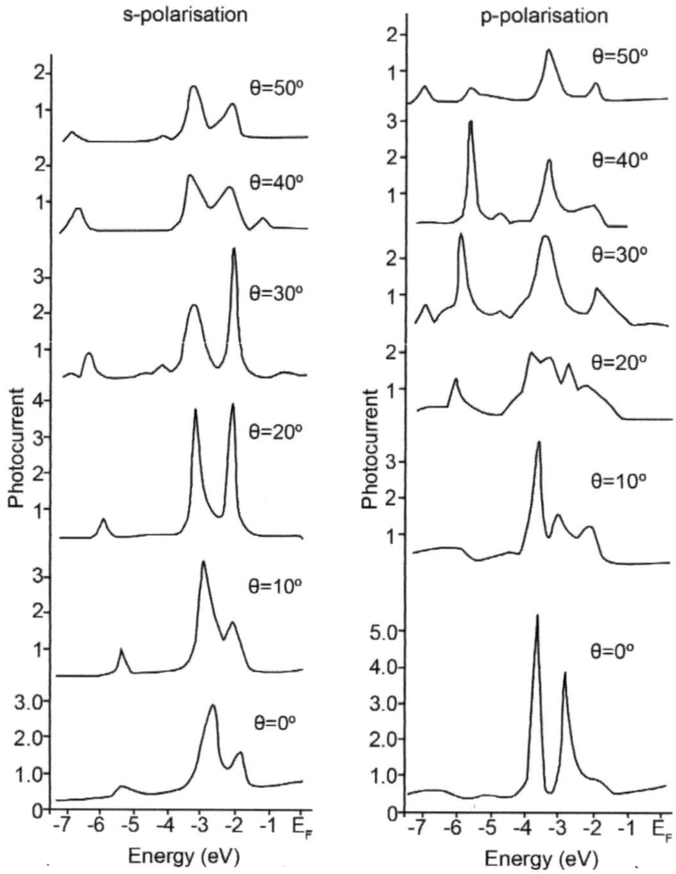

Figure 5: Calculated spectra 0 to 50 degrees: s and p polarization

1.6 Surface State

As we have discussed in the discussion for initial state calculation, the presence of the surface destroys the three dimensional periodicity. Since $k_\perp$ can be complex, it may be possible for some energy E that the matching

$$\phi_\alpha = \sum_g \phi_{g\alpha}(z)\exp i(k_{\|}.r_{\|}) \tag{7}$$

$$\psi k_{\|} = \sum_g T_g \frac{u_g(z)}{u_g(0)} \exp i(k_{\|}.r_{\|}) \tag{8}$$

will be satisfied for some complex $k_\perp$ with the wavefunction localised in the surface region. A schematic picture of such a wavefunction in one dimension is given below.

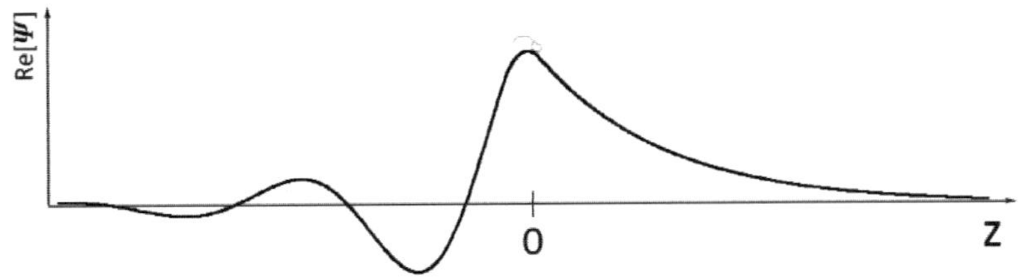

Figure 6: S chematic surface state wavefunction in 1 dimension

Such states are called surface states -they would exist in a band gap. If there are band states at the same energy then they would mix with the surface state, unless there are symmetry restrictions and these states are then known as surface resonances. However, the density of states would have a peak in the surface region. Photoemission is an excellent probe for detecting such surface states/resonances. The mean free path of electrons due to inelastic scattering in metals is of the order of 10 - 20 angstroms for the electron energy in the range of 10 - 100 eV (see e.g. [2] and [4]). So for photon frequency in the ultraviolet range, the electrons that are photoemitted come from a few layers in the surface region - the surface state/resonance effects do not get swamped by the bulk data.

Historically, the first surface state was observed by Field Emission on (100) face of tungsten about 0.4 eV below the Fermi energy (see [11] for details). Weng, Plummer and Gustafsson [12] showed that such states could also observed by angle resolved photoemission experiments on tungsten and molybdenum. The peaks due to such states go away if the clean surface has any contaminant and their position with respect to the Fermi energy is independent of the photon energy. We show their data below.

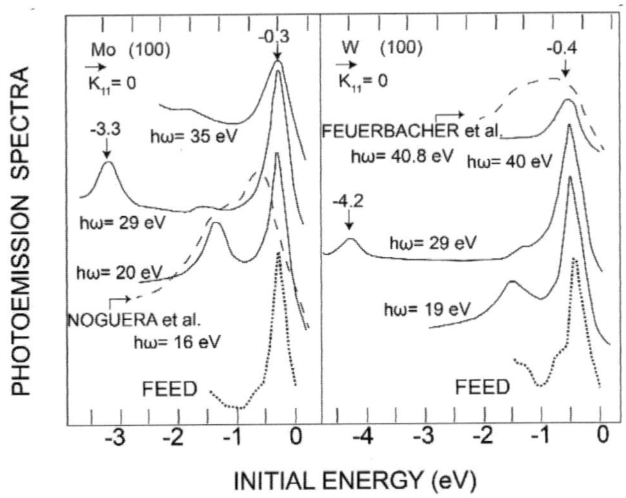

Figure 7: Surface states on W(100) and Mo(100) faces

The theoretical calculation of such surface states is a problem which needs accurate electronic state calculations in the presence of surfaces. Inglesfield [13] has given a very good review of surface electronic structure calculation. Such calculations can also be done in the framework of density functional theory by taking a layer by layer approach.

2. Spatial Variation of Photon Field

So far the spatial variation of the photon field has been completely neglected. However, photoemission experiments by Weng, Gustafsson and Plummer [14] on the surface state of tungsten(100) face, plotted against photon energy showed that the peak strength goes through a minimum near the bulk plasmon energy. This observed suppression in normal photoemission seems to be due to field variation in the surface region. In the context of jellium model Feibelman [15] calculated non-local response function and the electromagnetic field in the surface region and applied his results to free electron metals. Apart from some problems with photoemision results from aluminium his calculations were (i) too complex to be incorporated into photoemission programs and (ii) the jellium model could not be applied to transition metals or semiconductors. Bagchi and Kar [16] used a simple model - linearly interpolating the (local) dielectric function in the surface region from the bulk value (experimentally determined) and the vacuum value (unity). Taking the nominal surface plane to be at z = 0, the dielectric function in the surface region ($-a/2 \leq z \leq a/2$) can be written as

$$\varepsilon(\omega, z) = [1 + \varepsilon(\omega)]/2 + [1 + \varepsilon(\omega)](z/a)$$

This is a very simplistic model but the nice feature of this model is that the vector potential $\mathbf{A}_{\omega}$ (z) can be calculated analytically. For the wavefunctions a gaussian function was used for the surface state and free electron functions for the final state by Bagchi and Kar. The results showed that there was a valley in the plasma frequency region, in accordance with observed results. Thapa, Das and Kar used the model with Kronig-Penney and free electron wavefunctions and applied it to different materials to see the effect of spatially varying photon field (see [17] and [18]). Thapa and his co-workers are still using this model Currently, Thapa and his student Rebecca are do- ing calculations with this model and combining it with density functional method for calculating and also wavefunctions with muffin tin potentials with 2 dimensional periodicity. Let me conclude by showing a photoemission calculation from Boron Arsenide done by Rebecca and Thapa[19].

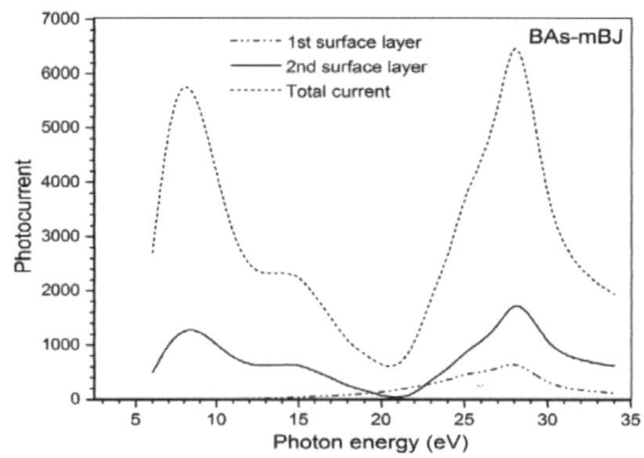

Figure 8: Calculated spectra from BAs

3. Conclusion

Angle Resolved Photoemission can give a wealth of data for material starting from clean and adsorbate covered metals to semi-conductors to cuprate superconductors. With the availability of synchrotron radiation, one can use a large range of photon frequencies. Both bulk electronic states as well as surface states and resonances can be probed. However, a complete theory of photoemission with all the quantities treated from first principles is not available. But focussing on appropriate choice of data one can make judicious approximations to theoretically interpret the experimental results.

Acknowledgment
Figures 1 and 6 are downloaded from Wikipedia and used here. Some old figures have been cleaned up with the help of Nandini Kar. The author is grateful to Prof. R. K. Thapa and the Organising Committee of the Condensed Matter Days 2016 (CMDAYS-2016) for the invitation to attend the conference and present the talk on which this article is based.

References

[1] Damascelli A, Hussain Z and Shen Z-X 2003 *Rev. Mod. Phys.* **75** 473
[2] Damascelli A 2004 *Physica Scripta* **T 109** 61
[3] Pendry J B *Photoemission and the Electronic Properties of Matter*, ed. B. Feurbacher, B. Fitton and R. Willis, p 87 Wiley, New York; 1978)
[4] Williams R H, Srivastava G P and McGovern 1980 I T *Rep. Prog. Phys.* **43** 1357
[5] Pendry J B 1974 Low Energy Electron Diffraction (Academic Press, New York)
[6] Liebsch A 1974 *Phys. Rev. Lettrs.* 32 1204
[7] Pendry J B 1975 *J. Phys. C: Solid St. Phys.* 8 2413
[8] Hopkinson J F L, Pendry J B and Titterington D J 1980 *Comput. Phys. Commun.* **19** 69
[9] Kar N 1981 *J. Phys.* **C 14** 2185
[10] Richardson N V and Sass J K 1981 *Surf. Sc.* **103** 496
[11] Gadzuk J W and Plummer E W 1973 *Rev. Mod. Phys.* **45** 487
[12] Weng S L, Plummer E W and Gustafsson T 1975 *Phys. Rev. B* **18** 1718
[13] Inglesfield J E 1982 *Rep. Prog. Phys.* **45** 223
[14] Weng S L, Gustafsson T and Plummer E.W. 1977 Phys. Rev. Lett **39** 822
[15] Feibelman P J 1975 *Phys. Rev. Lett.* **34** 1092
[16] Bagchi A and Kar N 1978 *Phys. Rev.* **B18** 5240
[17] Das P, Thapa R. K. and Kar N 1991 *Mod. Phys. Lett B* **5** 65
[18] Thapa R K, Das P and Kar N 1994 *Mod. Phys. Lett. B* **8** 361
[19] Lalngaihawmi R and Thapa R. K, unpublished data.

Investigation of doped Perovskite systems $RAlO_3$ using density functional theory based electronic structure and thermoelectric studies

Sandeep[1]*, D. P. Rai[2], A Shankar[3], M. P. Ghimire[1] and R. K. Thapa[1]

[1]Condensed Matter Theory Research Group, Physics Department, Mizoram University-796004, India
[2]Department of Physics, Pachhunga University College, Aizawl-796001, India
[3]Department of Physics, University of North Bengal, Darjeeling -734013

*Email: sndp.chettri@gmail.com

Abstract. Samarium doping effects on the thermoelectric properties in $Eu_{1-x}Sm_xAlO_3$ (x=0%, 50%, and 100%) were studied using first principles calculations based thermal transport property measurement. The result indicate that the compound is an intrinsic n-type material. Samarium doping has a positive effect on the overall thermoelectric performance of the $Eu_{1-x}Sm_xAlO_3$ system, with sharp increase in figure of merit (*ZT*) observed when x=0, 50 and 100% up to 150K. Compared to x=0 and 100%, the case of x=50% was found to have more positive increment in *ZT* value suggesting that the doing to have positive effect on figure of merit in $Eu_{1-x}Sm_xAlO_3$. Furthermore, all the samples show stable thermoelectric compatibility factors over a broad temperature range from 700 to 1000 K, which could have great benefits for their practical applications. It is concluded that the overall thermoelectric performance of the $Eu_{1-x}Sm_xAlO_3$ could be highly enhanced using doping techniques.

1. Introduction

Thermoelectric compounds have drawn renewed interest in the recent past due to their wide applicability in the field of device and high temperature components [1]. The perovskites have been known for their potential applications as electronic and thermoelectric devices [2-3] due to the rich exhibition of phase transformation with temperature and doping [4]. $EuAlO_3$ have been studied using mBJ and GGA+U based approximation to compute the electronic structures and were found to have high density of states at E_F supporting its candidature as thermoelectric device fabrication [5]. Thermoelectric power and electrical conductivity of $LaAl_{0.95}Mg_{0.05}O_{3-\delta}$ and $ErAl_{0.95}Mg_{0.05}O_{3-\delta}$ have been studied as a function of oxygen partial pressure [6]. The isobaric heat capacity of $SmAlO_3$ in the range 0-300 K has been measured with adiabatic calorimetry and its dependence on the electronic structures were reported with emphasis on 4*f* state electrons [7]. Thermo-electric efficient materials in pure or doped form have attracted the researchers in the last few decades, because of their striking features like easy synthesis, good conductance, price and stability [8]. Wilfried *et al.* studied the large Seebeck coefficient and its dependence on effective electron mass and suggested doping and increase of lattice constant [9]

2. Methodology

Seebeck coefficient (S), thermal conductivity (κ) and electrical conductivities (σ) for an isotropic electronic structure using BoltzTrap code [10]. Eq. (1) interprets electrical conductivity tensors [11]

$$\sigma_{\alpha,\beta} = e^2 \sum_{i,k} \left[\frac{-\partial f_0(T,\varepsilon,\mu)}{\partial \varepsilon}\right] v_\alpha v_\beta \tau_k \qquad (1)$$

where, α, β are the tensor indices, $v_\alpha v_\beta$ are the group velocities, e is the electron charge and τ_k is the relaxation time. The relaxation time τ is neglected for isotropic systems as the variation of τ is not known. The electrons contribution remain near the chemical potential (μ) in a narrow range of $\mu - k_B T < \varepsilon < \mu + k_B T$, where k_B is the Boltzmann constant [12]. The transport distribution is written as [13]

$$\Xi_{i,k} = e^2 \sum_{i,k} v_\alpha v_\beta \tau_k \qquad (2)$$

which is the kernel of all transport coefficients. From the rigid band approach, the electrical conductivity, thermal conductivity and Seebeck coefficient can be written as a function of temperature (T) and chemical potential (μ) by integrating the transport distribution [10, 14]

$$\sigma = \frac{1}{\Omega} \int \Xi_{i,k} \left[\frac{-\partial f_0(T,\varepsilon)}{\partial \varepsilon}\right] d\varepsilon \qquad (3)$$

$$\kappa = \frac{1}{e^2} k_B^2 T \int \Xi_{i,k} \left(\frac{\varepsilon-\mu}{k_B T}\right)^2 \left[\frac{-\partial f_0(T,\varepsilon)}{\partial \varepsilon}\right] d\varepsilon \qquad (4)$$

$$v_{\alpha\beta} = \frac{k_B}{e\Omega} \int \Xi_{i,k} \left(\frac{\varepsilon-\mu}{k_B T}\right) \left[\frac{-\partial f_0(T,\varepsilon)}{\partial \varepsilon}\right] d\varepsilon \qquad (5)$$

Here f_0 is a Fermi-Dirac distribution function and Ω is the volume of a unit cell. The Seebeck coefficient (S) is determined as

$$S_{ij} = E_i(\nabla_j T)^{-1} = (\sigma^{-1})\alpha_i v_{\alpha j} \qquad (6)$$

The thermoelectric efficiency is presented as

$$ZT = \frac{S^2 \sigma T}{\kappa} \qquad (7)$$

Thus the efficiency of thermoelectric materials (ZT) can be tuned by optimizing any of the three parameters S, σ and κ at defined temperature T.

3. Results and discussions

The electronic structure studies have been performed by Sandeep *et al.* [5] using both mBJ as well as GGA+U approximations. The contribution of density of states in the vicinity of Fermi level (E_F) motivated the thermoelectric studies of doped $EuAlO_3$. The negative value of S indicates n-type carriers (Fig. 1a). Since S depends on temperature, the sharp decrease of S at lowest temperature is due to the depopulation of phonon modes.

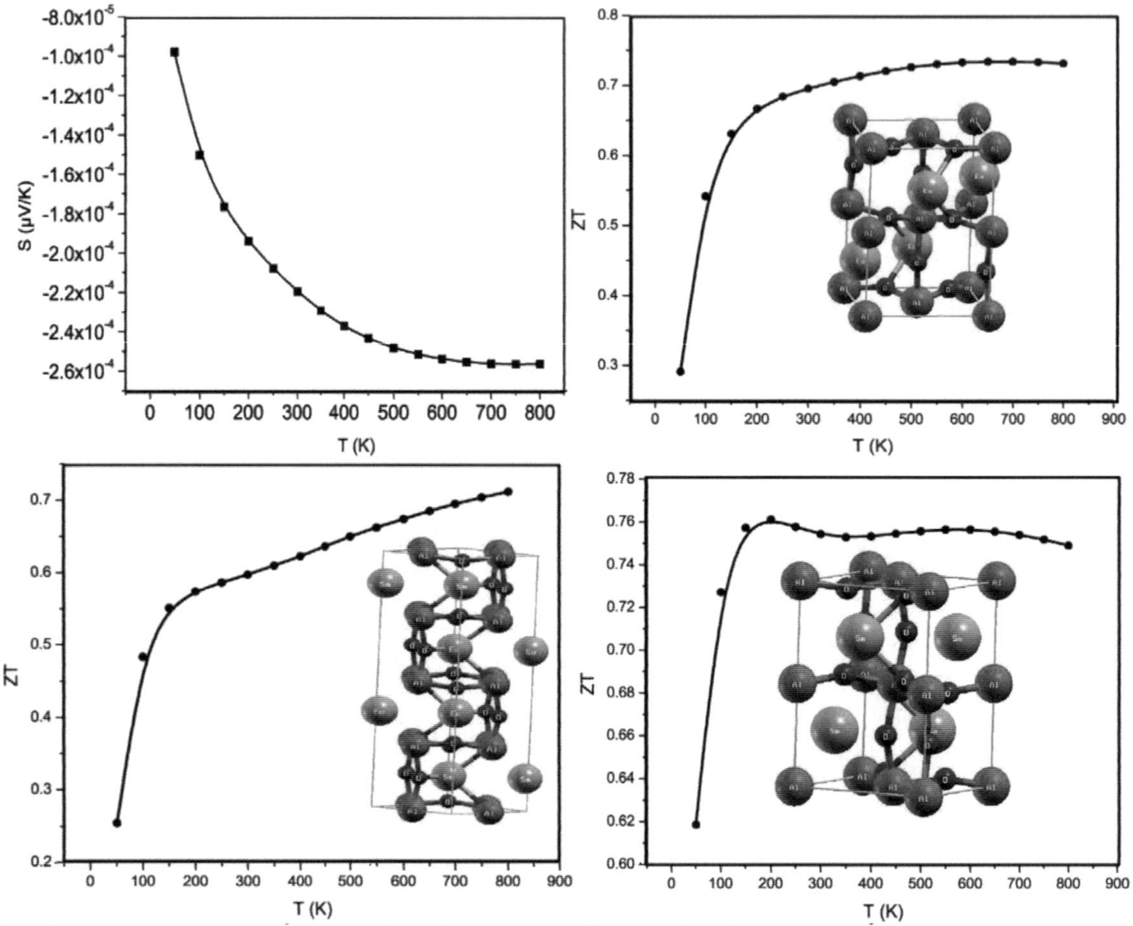

Fig 1. Temperature dependence of (a) Seebeck coefficient (S); Temperature dependence of figure of merit (ZT) calculated within GGA for $Eu_{1-x}Sm_xAlO_3$ with (b) x=0%, (c) x=50% and (d) x=100%.

Strongly correlated heavy Fermion compounds of Ce and Yb also show similar type of low temperature phenomena [15]. The value of S at room temperature was found to be -2.19 ×10 -4 μV/K. The figure of merits are shown in Fig 1(b-d) for three different doping concentrations along with their structures for x=0%, 50% and 100% respectively. It was found that below 150 K all the three ZT values for different doping states were of similar nature. They showed sharp increment with temperature below 200 K. x=50% doping configuration was found to have a more positive effect of temperature on it compared to x= 0% and 50% suggesting the presence of doped Sm in $EuAlO_3$ enhanced the thermoelectric efficiency of the material and is also indication that a doping of other such concentration may yield higher ZT values which may be useful for thermoelectric device fabrications.

Acknowledgment
SD acknowledges DST-SERB(India) FAST Track project project.

References
[1] Di Salvo F J 1999 *Science* **285** 703-706
[2] Sandeep, Rai D P, Shankar A, Ghimire M P, Khenata R and Thapa R K 2016 *Mod. Phys. Lett. B*, 2883-2887
[3] Hashimoto H, Kusunose T and Sekino T 2011 *Journal of Ceramic Processing Research* **12,** 223-227

[4] Karvonen L, Tomes P and Weidenkaff A 2011 *Material Matters* **6** 92–99
[5] Sandeep, Rai D P, Shankar A, Ghimire M P, Khenata R and Thapa R K 2015 *Phys. Scr.* **90** 065803-065810
[6] Ahlgren E O, Ranlov J and Poulsen F W 1995 *J. Electrochem. Soc.* **142** 4230-4234
[7] Kopan A R, Gorbachuk M. P, Lakiza S. M. and Tishchenko Ya S 2012 *Powder Metallurgy and Metal Ceramics* **51** 209-216
[8] Hochbaum A I, Chen R, Delgado R D, Liang W, Garnett E C, Najarian M, Majumdar A and Yang P, 2008 *Nature* **451** 163-167
[9] Wunderlich W, Ohta S, Ohta H, Koumoto K, "Effective mass and thermoelectric properties of $SrTiO_3$-based natural superlattices evaluated by ab-initio calculations" in 24th International Conference On Thermoelectrics, (ICT 2005 IEEE Proceedings) pp. 237-240
[10] Madsen G K H, Singh D 2006 *Comput. Phys. Commun* **175** 67-71
[11] Yamamoto A, Miyazaki H, Takeuchi T 2014 *J. Appl. Phys.* **115** 023708-023717
[12] Takeuchi T 2009 *Materials Transactions* **50** 2359-2365
[13] Mahan G D and Sofo J O 1996 Proc. Natl. Acad. Sci. USA **93** 7436-7439
[14] Scheidementel T J, Draxl C A, Thoonhauser T, Badding J V, Sofo J O 2003 *Phys. Rev. B* **68** 1252101-1252106
[15] Dilley N R, Bauer E D, Maple M B, Dordevic S, Basov D N, Freibert F, Darling T W, Migliori A,.C. Chakoumakos B, Sales B C 2000 *Phys. Rev. B* **61** 4608-4614

Basic concepts of Density Functional Theory: Electronic structure calculation

B. Indrajit Sharma[1] and R K Thapa[2]

[1]Department of Physics, Assam University Silchar, Assam, India
[2]Department of Physics, Mizoram University, Aizawl, Mizoram, India

*Email: indraofficial@rediffmail.com

Abstract. We are looking for a material which possesses the required properties as demanded for technological applications. For this we have to repeat the preparation of the appropriate materials and its characterizations. So, before proceeding to experiments, one can study on computer generated structure and predict the properties of the desired material. To do this, a concept of Density Functional Theory comes out.

1. Introduction

The first principles approach to condensed matter theory is widely used. It starts from the very basic principles that materials are made of atoms. These atoms are again made of a positively charged nucleus and a number of negatively charged electrons. The interactions between atoms are determined by the interactions of the constituent electrons and nuclei that form the basic interaction. Thus if we can model interactions accurately, then most of the complex physical properties can be understood. Thus the *ab initio* calculation or first-principles calculation do not depend upon any external parameters except the atomic numbers of the constituent atoms to be modelled and cannot therefore be biased by preconceptions about the final result. Such calculations are reliable and can be used with confidence to predict the behaviour of nature.

2. Concepts of Density Functional Theory:

Density functional theory (DFT) is the basic theory of electronic structure calculations in Condensed Matter Physics. It was so challenge to approximate functional and computational cost. In 1990's there are widely used functional that give a quite reasonable accuracy with the affordable computational cost. In simple language, Density functional theory is a completely different in terms of its approaching any interacting problem that by mapping it exactly to a non-interacting problem that gives an alternative way to tackle the problem. Thus DFT is a ground-state theory in which the charge density is the relevant physical quantity. DFT is very successful in describing structural and electronic properties in materials science and chemical science.

Density functional theory is developed based on the Hohenberg-Kohn theorem. There are two parts in Hohenberg-Kohn Theorem.

Hohenberg-Kohn Theorem –I:

This theorem states that the external potential V_{ext}(r) is determined (within a trivial additive constant) by the electron density ρ(r).

It implies that for a given density there will be a single V_{ext}(r). In other words, there cannot be two V_{ext}(r) for a given ρ(r). Thus, this theorem demonstrates that the electron density specifies uniquely the Hamiltonian operator.

In consequence, an alternative statement of the first Hohenberg Kohn theorem is that any ground state expectation value corresponding to an observable $\hat{O}$, is a functional of the density,

$$O[\rho] = \left\langle \psi[\rho] \middle| \hat{O} \middle| \psi[\rho] \right\rangle$$

Hohenberg-Kohn Theorem –II:

The second Hohenberg Kohn theorem states that the functional $F_{HK}[\rho]$ that delivers the ground state energy if and only if the input density is the true ground state density.

In other words,

For a trial density $\tilde{\rho}(r)$ such that $\tilde{\rho}(r) \geq 0$ and $\int \tilde{\rho}(r)dr = N$

$$E[\rho] \leq E[\tilde{\rho}]$$

where $E[\tilde{\rho}]$ is the energy functional.

$$E[\tilde{\rho}] = \int \tilde{\rho}(r)V_{ext}(r)dr + F_{HK}[\tilde{\rho}]$$

2.1 Kohn-Sham equations

We know that a solid consists of heavy positively charged particles (nuclei) and lighter negatively charged particles (electrons). The nuclei are much heavier and therefore much slower than the electrons. We can consider the nuclei are at fixed positions. So, the nuclei become the external potential to the electrons. This is the Born-Oppenheimer approximation. In the further discussion, we use the atomic units (that is $e = m_e = \hbar = 1$; $4\pi\epsilon_o=1$). Now, after Born-Oppenheimer approximation, we can see the Solid is generally a system of interacting electrons in the field of fixed ions described by the Hamiltonian,

$$\hat{H} = \sum_{i=1}^{N}\left(-\frac{1}{2}\nabla_i^2\right) + \sum_{i=1}^{N} V_{ext}(r_i) + \sum_{i<j}^{N}\frac{1}{r_{ij}} \tag{1}$$

where "r_{ij}" is the separation between i^{th} and j^{th} electrons;

$V_{ext}(r_i) = -\sum_{I=1}^{M}\frac{Z_I}{r_{iI}}$ is the external potential acting on i^{th} electron due to the M nuclei and "r_{iI}" is the separation of ith electron and I^{th} nucleus.

A stationary electronic state is then described by a wave function $\Psi(r_1, r_2, ...r_i......r_N)$ satisfying the many electron Schrödinger equation,

$$\hat{H}\psi = E\,\psi$$

that is

$$[\hat{T} + \hat{V}_{ext} + \hat{V}_{ee}]\psi = E\psi \tag{2}$$

Hohenberg and Kohn proved that the total energy of a system in the presence of the static external potential is unique functional of the charge density.

$$E[\rho]=T[\rho]+V_{ee}[\rho]+V_{ext}[\rho]$$
$$E[\rho]=F_{\text{HK}}[\rho]+\int\rho(r)V_{ext}(r)d^3r \tag{3}$$

Here the term,

$T[\rho]+V_{ee}[\rho]=F_{HK}[\rho]$

is same for all solid and thus it is universal. This does not depend on the particular system under consideration and is called Hohenberg-Kohn functional.

In the non-interacting case, $V_{ee}=0$,

Thus the ground state energy $E[\rho]$ has a kinetic contribution and a contribution from the external potential.

$$E[\rho]=T_o[\rho]+V_{ext}[\rho]=T_o[\rho]+\int\rho(r)V_{ext}(r)d^3r \tag{4}$$

Again, the ground state wave function of the non-interacting system can be written as a Slater determinant with the orbitals satisfying the single particle Schrodinger equation,

$$\left[-\frac{1}{2}\nabla^2+V_{ext}(r)\right]\phi_m^o(r)=\varepsilon_m^o\phi_m^o \tag{5}$$

The ground state density is then given by a sum over occupied states,

$$\rho(r)=\sum_m^{occup}\left|\phi_m^o(r)\right|^2 \tag{6}$$

where the orbital ϕ_m^0 are normalized so that the density satisfies the correct condition to the number of particles N.

Thus, we can write the ground state energy for this non-interacting case as

$$\sum_m^{occup}\varepsilon_m^o=T_o[\rho]+\int\rho(r)V_{ext}(r)d^3r \tag{7}$$

Now, for the interacting case, we can represent the energy functional $E[\rho]$ for a many electron system in the form,

$$E[\rho]=T_O[\rho]+\int\rho(r)V_{ext}(r)d^3r+\frac{1}{2}\iint\frac{\rho(r_1)\rho(r_2)}{|r_1-r_2|}d^3r_1d^3r_2+E_{XC}[\rho(r)] \tag{8}$$

Where

$\frac{1}{2}\iint\frac{\rho(r_1)\rho(r_2)}{|r_1-r_2|}d^3r_1d^3r_2=V_H[\rho]$ is the Hatree energy

and $E_{XC}[\rho(r)]$ is the exchange correlation energy that contains all the contributions not taken into account in the kinetic energy functional of non-interacting electron gas, the external and the Hatree energy respectively.

Thus, we can write the analogous equation [equation (5) for non-interacting case] for the interacting electron system as

$$\left[-\frac{1}{2}\nabla^2+V_{eff}(r)\right]\phi_m(r)=\varepsilon_m\phi_m \tag{9}$$

where the single particle wave functions $\phi_m(r)$ are the N lowest energy solutions of the above equation (9) called the Kohn Sham equations and $\phi_m(r)$ are the Kohn Sham orbital that reproduce ρ(r) of the original many-body problem.

$$V_{eff} = V_{ext}(r) + \int \frac{\rho(r^{/})}{|r - r^{/}|} d^3 r^{/} + \frac{\delta E_{XC}[\rho]}{\delta \rho(r)}$$

That is

$$V_{eff} = V_{ext}(r) + V_H(r) + V_{XC}(r) \tag{10}$$

where

$V_H(r) = \int \frac{\rho(r^{/})}{|r - r^{/}|} d^3 r^{/}$ is the Hatree potential.

$V_{XC}(r) = \frac{\delta E_{XC}[\rho]}{\delta \rho(r)}$ is the exchange correlation potential.

2.2 Easy Steps for Computational DFT

We recall the representation of total energy [equation (8)] of the interacting electrons system,

$$E[\rho] = T_o[\rho] + \int \rho(r) V_{ext}(r) d^3 r + \frac{1}{2} \iint \frac{\rho(r_1)\rho(r_2)}{|r_1 - r_2|} d^3 r_1 d^3 r_2 + E_{XC}[\rho(r)]$$

So, we have from equation (10),

$$V_{eff} = V_{ext}(r) + V_H(r) + V_{XC}(r)$$

$$V_{ext} = V_{eff}(r) - V_{XC}(r) - V_H(r)$$

Incorporating this expression in the above equation, we can write as

$$E[\rho] = \sum_{m}^{occup} \varepsilon_m^o - \frac{1}{2} \iint \frac{\rho(r_1)\rho(r_2)}{|r_1 - r_2|} d^3 r_1 d^3 r_2 - \int V_{XC}(r) \rho(r) d^3 r + E_{XC}[\rho(r)] \tag{11}$$

Equation (11) gives the final expression for the ground state energy of the system of interacting electrons. This is the expression used in computational calculation to determine the ground state energy functional for the interacting system of electron.

References:

[1] Tran F and Blaha, P *Phys. Rev. Lett* **102** (2009) 226401.
[2] Perdew J P, Burke S and Ernzerhof M, *Phys. Rev. Let.* **77** (1996) 3865.
[3] Hohenberg P and Kohn W, *Phys. Rev.* 3B **136** (1964) 864.
[4] Kohn W and Sham L J, *Phys. Rev.* 4A **140** (1965) 1133.
[5] Cottenier S, *Density Functional Theory and the family of (L)APW-methods: a step-by-step Introduction*, ISBN90-807215-1-4 (2002).
[6] Blaha P, Schwarz K, Madsen G K H, Kvasnicka D and Luitz J, *WIEN2k An Augmented Plane Wave + Local Orbitals Program for Calculating Crystal Properties*, ISBN3- 9501031-1-2 (2001)

Electronic and Piezoelectric properties of half-Heusler compounds: A first principles study

D. P. Rai[1]*, Sandeep[2], A. Shankar[3], Abeer E. Aly[4], P. K. Patra[5] and R. K. Thapa[2]
[1]Department of Physics, Pachhunga University College, Aizawl, India-796001,
[2]Department of Physics, Mizoram University, Aizawl, India-796009
[3]Department of Physics, North Bengal University, Siliguri, Darjeeling, India-734013
[4]Faculty of information System and Computer Science , October 6 University Cairo, Egypt.
[5]Faculty in Centre for Science Education, NEHU Shillong, India-794022.

E-mail: dibyaprakashrai@gmail.com

Abstract. We have investigated the semiconducting and piezoelectric properties of bulk MNiSn (M=Ti, Zr, Hf) type a half-Heusler compound with cubic F-43m symmetry by means of density functional theory (DFT). For electron exchange correlation a generalized gradient approximation (GGA) was used. Special attention was paid to establish a most favourble ground state configuration on magnetic as well as non-magnetic ordering. With fully optimized structure the electronic and ferroelectric calculation was performed. The formation of band gap was discussed on the basis of d-d orbital hybridization. Further we have calculated the spontaneous polarization by means of structural deformation.

1. Introduction

The rapid environmental hazards due to the over consumption of fossil fuels and depletion of natural energy resources is a great threat to the modern world. In order to tackle problems of energy crisis, there arises a critical need of innovative and eco-friendly technologies based on multi-functional materials. Nevertheless, the current researches are devoted in the exploitation of material's properties for technological application or designing of such kind of materials that act smartly in response to external parameters like temperature, field and stresses. The materials are to be design in such a way to obtain high speed, high accuracy, low power consumption, improved communications and yield high energy conversion/storage. Until now different variety of materials are studied from experiment and theoretical approach (first principles calculation) and diverse functional properties were reported. Such as Chalcopyrites for solar cells [1], Heusler compounds for spintronic [2-5], Skutterudite for thermoelectric [6], Perovskite for solar cell, ferroelectric[7,8], etc. Among all other Heusler compounds a half-Heusler compounds are of great interest as they crystallizes in $C1_b$ cubic structure with space group F-43m and possess high curie temperature [9]. Also the half-Heusler compounds with 18 total valence electrons are non-magnetic semiconductors with band gap at the Fermi level. The formation of band gap is due to the hybridization between M-d and Ni-d orbitals [10]. Possessing semiconductor characteristic with no inversion center in MNiSn makes as it a potential candidate for ferroelectric/piezoelectric materials. Therefore, we employed a first-principles calculation to explore the microscopic origins of the functional properties of interest, and to obtain the information regarding the ferroelectric/piezoelectric properties by deforming the crystal structures.

[1] To whom any correspondence should be addressed.

Journal of Physics: Conference Series **765** (2016) 012005 doi:10.1088/1742-6596/765/1/012005

2. Computational Detail

The crystal structure of MNiSn is shown in Fig. 1. In this work, we have employed the FP-LAPW method within the framework of the density functional theory (DFT) [11] as implemented in the WIEN2k code [12]. We have also used another code called BerryPI an interface with WIEN2k to calculate the piezoelectric properties. Experimental lattice constants 6.11 A° and 5.94 A° [13] were used for volume optimization to obtain optimized lattice constants 6.23 A° and 6.06 A° for ZrNiSn and TiNiSn, respectively.

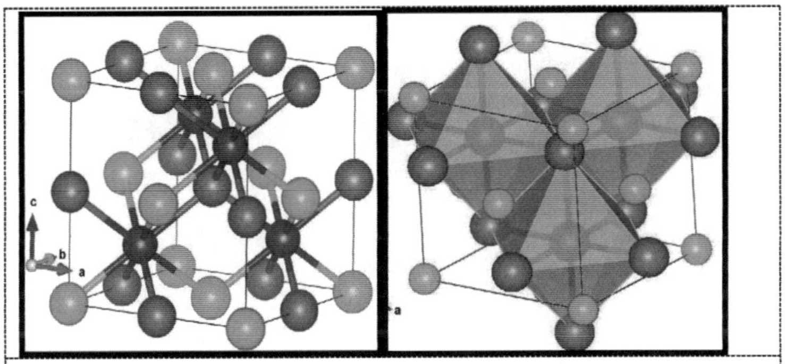

Figure 1. Crystal structure of MNiSn hal-Heusler compounds with F-43m cubic symmetry (M-Green, Ni-Red & Sn-Blue).

3. Results and Discussion

In this section we have discussed the electronic and piezoelectric properties of MNiSn, a half-Heusler compounds. The calculated electronic structures of MNiSn is presented in the form of density of states, as shown in Fig. 3(a-d). Figure 3(a) shows the result of total DOS, as we have seen the energy gap is located exactly at the Fermi level, which is a signature of semiconducting characteristic. The formation of band gap in MNiSn compounds with 18 valence electrons are described by Galanakis *et al.* [10] on the basis of d-d hybridization, as shown in Fig. 2. Fig. 3(a-d) displayed the DOS of MNiSn with majority contribution from Ni-d states in the valence region and M-d states contributes maximum in the conduction region. The hybridization of d-d orbitals take place near the Fermi energy. The bondings are formed between the M (d-t_{2g}) and Ni (d-t_{2g}) also between M(d-e_g) and Ni(d-e_g) as a result the energy level shifts down in the valence region. Corresponding anti-bondings are formed between M(d-e_g/d-t_{2g}) and Ni(d-e_g/d-t_{2g}) which gives an unstable state, and shifts the energy level high above the Fermi energy in the conduction region. Thus, the measure of energy difference between the highest occupied band and the lowest unoccupied gives the value of energy gap. Our calculated band gaps are in good agreement with the previous results [4, 14, 15, 16].

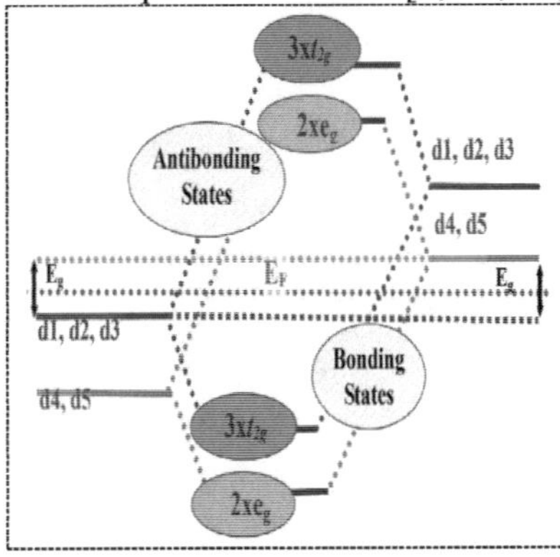

Figure 2. A schematic diagram shows the hybridization of d-d orbitals with the opening of the band gap. As we know d-orbitals are degenerated into 5 states 3 d-t_{2g} and 2 d-e_g states.

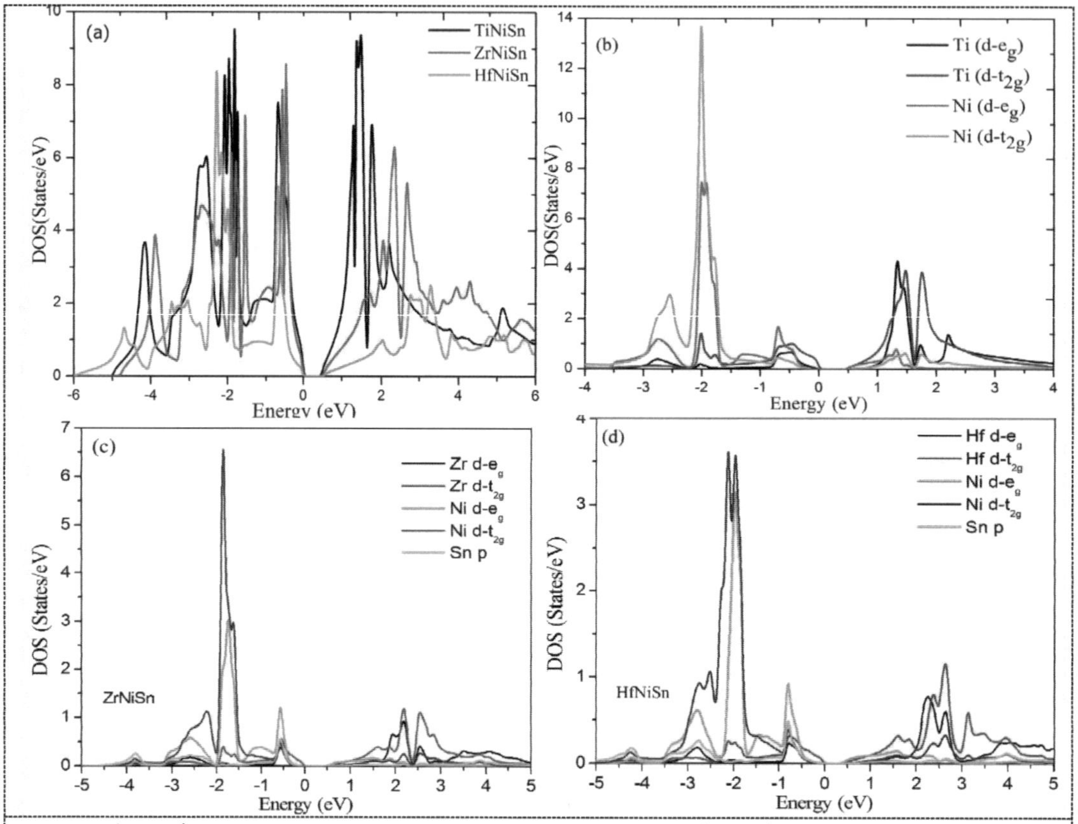

Figure 3. (a) Total DOS, partial DOS of (b) TiNiSn, (c) ZrNiSn and (d) HfNiSn

The presence of band gap in MNiSn predicts the semiconducting behaviour along with no inversion center. As we know these type of compounds are very interesting and may give some functional properties like ferroelectricity/piezoelectricity. The calculation of ionic contribution is based on the position of atomic nuclei and corresponding ionic charges. Hence, the piezoelectric properties are calculated by deforming the structures along the z-axis with shreaing stress C_{44}. The piezoelectric coefficient '*P*' is calculated by $P = P_{ion} + P_{el}$, here P_{ion} is the polarization due to ionic displacement and P_{el} is the electronic polarization. The complete equation of total polarization is given as [17]

$$P = \frac{e}{\Omega}\sum_{s}^{atoms} Z_s^{ion} r_s - \frac{2ei}{(2\pi)^3}\int_{BZ} dk < u_{nk}\left|\widetilde{N}_k\right|u_{nk} > \qquad (1)$$

where Ω is the simulation cell volume, 'e' is the elementary charge, Z_s^{ion} is the ionic charge represented by the number of valence electrons in the atom 's' and r_s is its position vector. The factor of 2 in the numerator of second term corresponds to the band occupancy. The integration of second term in Eq.(1) is performed over the Brillouin zone (BZ) and the integrand is closely related to the geometrical phase change [18]. The piezoelectric constant is calculated as $d = P/C_{44}$. The calculated values of piezoelectric parameters are presented in Table 1.

Journal of Physics: Conference Series **765** (2016) 012005 doi:10.1088/1742-6596/765/1/012005

Table 1. *a* is lattice constant, E_g is band gap, *P* is piezoelectric coefficient, *d* is piezoelectric constant, C_{44} is shearing stress and ε_o is static dielectric constant. (*Ref. [11])

MNiSn	a (A^o)	E_g (eV)	P (C/m^2)	C_{44} (GPa)	d (pC/N)	ε_o
TiNiSn	5.94*/6.06	0.44*	0.59/0.81*	67.78/62.0*	8.78/12.9*	27.94/30*
ZrNiSn	6.11*/6.23	0.52*	0.16/0.19*	77.21/70.0*	2.0/2.8*	24.57/26*
HfNiSn	6.08*/6.13	0.41*	0.19/0.28*	80.77/80.0*	2.4/3.4*	21.74/24*

4. Conclusion.

We have calculated the electronic structure and piezoelectric properties of MNiSn (M=Ti, Zr, Hf) within FP-LAPW method. Our calculated results of electronic band gaps are in good agreement with the experimental one. We have discussed the presence of 18 valence electrons and its importance in the formation of band gap with d-d orbital hybridization. The presence of band gap at the Fermi level predict our compounds MNiSn is a semiconductor. The compounds with semiconducting behaviour possess diverse functional properties, one of them is piezoelectricity. We have also calculated the piezoelectric parameters which are in good agreement with the previous data.

Acknowledgment

DPR acknowledges UGC Start-Up-Grant No.F.30-52/2014/BSR (New Delhi, India) and RKT a research project from SERB via EMR/2015/001407.

References

[1] Mise T, and Nakada T 2010 *Thin Solid Films* **518** 5604-5609.
[2] Rai D P, D. T. Khating, Patra P K, Hashemifar S J, Jamal M, Lalmuanpuia, Ghimire MP, Sandeep, Rosangliana and Thapa R K 2010 *Indian Journal of Physics* **84** 593.
[3] Rai D P, Sandeep, Shankar A, Ghimire MP and Thapa R K 2012 *Physica Scripta.* **86** PS422049.
[4] Rai D P, Shankar A, Sandeep, Ghimire MP, Khenata R and R. K. Thapa 2016 *RSC Adv.* **6** 13358.
[5] Rai D P, Sandeep, Shankar A, Ghimire MP, A P Sakhya A P, Sinha T P, Khenata R and R. K. Thapa 2016 *Materials Research Express* **3** 075022.
[6] Bravo M R, Moure A, Fernandez J F and Gonzalez M M 2015 *RSC Adv.* **5** 41653-41667.
[7] Collavini S, Volker S F and Delgado J L 2015 *Angew. Chem. Int. Ed.* **54** 9757-9759.
[8] Bendek N A, Rondivelli R N, Djani H, Ghosez P and Lightfoot P 2015 *Dalton Trans.* **44** 10543-10558.
[9] Wurmehl S, Fecher GH, Ksenofontov V, Casper F, Stumm U, Felser C, Lin H J and Hwu Y 2006 *J. Appl. Phys.* **99** 08J103.
[10] Galanakis I, Dederichs P H and Papanikolaou N 2002 *Phys. Rev.* **B 66** 174429.
[11] Kohn W and Sham L 1965 *J. Phys. Rev.***140** 1133.
[12] Blaha P, Schwarz K, Madsen G K H, Kvasnicka D, Luitz J and Schwarz K 2013 *An Augmented Plane Wave plus Local Orbitals Program for calculating Crystal Properties: Wien2k User's Guide(Wien: Technische Universitat Wien)* (2013).
[13] Roy A, Bennett J W, Rabe K M and Vanderbilt K *Phys. Rev. Lett.* 2002 **109** 037602.
[14] Aliev F G 1991 *Phys. B* **171** 199.
[15] Ogut S and Rabe K M 1995 *Phys. Rev. B: Condens. Matter Mater.Phys.* **51** 10443.
[16] Heyd J, Scuseria G. E and Ernzerhof M 2003 *J. Chem. Phys.* **118** 8207.
[17] Ahmed S J, Kivinen J, Zaporzan B, Curiel L, Pichardo S and Rubel O 2013 *Comput. Phys. Comm.* **184** 647-651 (2103).
[18] Berry M V 1984 *Proc. R. Soc. Ser.* **A 392** 45.

Bending analysis of carbon nanotubes using nonlocal continuum models

P K Patra
Centre for Science Education, North-Eastern Hill University, Shillong 793022, India
Email: pkpatranehu@gmail.com

Abstract. Experiments show that a CNT can withstand large bending curvature reversibly. As the deflection is severe for large varying forces, the elementary elastic theory cannot be applied. Again, the classical (or local) continuum elastic theory is questionable at very small length scale (less than 100 nm) because the material microstructure, such as lattice spacing between individual atoms, become important. For studying mechanical properties of carbon nanotubes (CNTs) of small length, the classical theory can be modified by using nonlocal continuum models proposed by Eringen. In the present work, static analysis of CNT is presented using a nonlinear relation between the bending curvature and the beam deflection and the nonlocal continuum mechanics models. It is shown that there is an upper bound on the load on a cantilevered CNT.
Keywords: Nonlocal continuum models, carbon nanotubes, static analysis, upper bound.

1. Introduction

Both experimental and theoretical studies show that CNTs exhibit superior mechanical, electronic, electrochemical and thermal properties over many known materials. Many experiments have been performed to study their mechanical properties like strength, stiffness etc. which can be inferred from measuring the values of tensile strength (σ_S) and Young's modulus (E) respectively. Experimental measurements of mechanical property of CNTs involve the techniques such as transmission electron microscope (TEM), scanning electron microscope (SEM) and atomic force microscope (AFM). From experiments it is found that CNTs have high tensile strength (~150 GPa) and Young's modulus (~ 1TPa) in addition to small size (~ nm) and low weight (~ one-sixth of steel). Because of these properties they present a promising nanoscale material for building different kinds of nanodevices in nanoelectromechanical systems (NEMS) and many other applications. Most potential applications of CNTs are heavily based on a thorough understanding of their mechanical behaviours. However, their mechanical properties have not fully predicted. A detail study of the mechanical properties of CNTs will be important for such applications. Most estimates of E have been obtained by using elementary elastic theory.

In order to determine the effective value of E of multiwalled carbon nanotubes (MWCNTs), in an experiment, Wong *et al.* [1] applied an external force at different locations along a cantilevered MWCNT and measured deflection versus bending force using AFM. They observed that the CNT show remarkable flexibility and large elastic deformation without breaking. And also they noted that the strain energy increases quadratically with deflection as expected for a harmonic system, whereas for large deflection the strain energy showed a deviation from the usual harmonic behaviour. Some more experiments have also shown that CNTs can withstand large bending curvatures reversibly [2-3]. The linear theory, which is valid only for small bending deformations, cannot explain such behaviour. However, the above behaviour can be explained by considering the nonlinear relation between the bending curvature $K(x)$ and the beam deflection $\omega(x)$ [4].

Carbon nanotubes are of nanometer size and the experimental error is quite high due to lack of information about its diameter and thickness of layers. Further, conducting experiments with such types of specimens are difficult and costly. Hence it is essential to develop an appropriate mathematical model to study its mechanical properties, which is the subject of recent research in computational nanomechanics and computational condensed matter physics. Two approaches are widely used for the theoretical study on CNTS: molecular dynamics simulation and continuum mechanics methods. However, the former one is very time-consuming and remains formidable for large-scale systems. Hence we will follow the latter in which CNTs are modelled as elastic beams such as Timoshenko, Euler-Bernoulli, shells and elastic rods, which have been widely and successfully used to study the mechanical behaviour of nanostructures such as CNTs, graphene sheets, nanofibres etc. Again, the classical continuum elastic theory, which is quite reliable for mechanical properties of CNTs of length scales 100 nm and more [5], is questionable at very small length scale (less than 100 nm) [6] because the material microstructure, such as lattice spacing between individual atoms, become important. This suggests that the classical theory needs to be modified. This was formally done by Eringen's nonlocal continuum models [7] which extend the continuum approach to smaller length scale.

In the present work, we have used nonlocal constitutive equations of Eringen [8] and the nonlinear relation between the bending curvature $K(x)$ and the beam deflection $\omega(x)$ to obtain analytic solution for large deflection for a cantilevered CNT. We have found that there is an upper bound on the load that can be applied at the tip of the free end of a cantilevered CNT which can be regarded as the maximum load needed for breaking the tube.

2. Fundamental aspects of non-local continuum models

The classical continuum models assume that the stress at a given point x in a body is dependent uniquely on the strain state at that same point in the body. Hence the models are usually called local continuum models and local elasticity is inherently size-independent. The constitutive equation of classical elasticity, an algebraic relationship between the stress and strain, is given by

$$\sigma(x) = E\,\epsilon(x), \tag{1}$$

where $\sigma(x)$ and $\epsilon(x)$ are the stress and the strain respectively. However, at nanometer scales the microstructure of the material, such as the lattice spacing between individual atoms, becomes increasingly important and the discrete structure of the material can no longer be homogenized into a continuum. The nonlocal elasticity models of Eringen [7] abandon the classical assumption of locality and assume that the stress at any reference point in the body depends not only on the strains at that point but also on the strains of at all points of the body. This is based on the atomic theory of lattice dynamics and some experimental observations on phonon dispersion. If the effects of strains at points other than the reference point are ignored, one obtains local continuum model. Thus in nonlocal continuum mechanics, the long range forces between atoms and the internal scale effect are considered in construction of constitutive equations. The most general form of the constitutive equation for nonlocal elasticity involves an integral over the entire region of interest. This integral contains a kernel function which describes the relative influences of strains at various locations on the stress at a given location.

For homogeneous and isotropic elastic solids, the nonlocal stress tensor at point x is given by

$$\sigma_{ij}(\boldsymbol{x}) = \int_v \alpha(|\boldsymbol{x}' - \boldsymbol{x}|, \tau)\, \sigma_{ij}^{(c)}(\boldsymbol{x}')\, dV(\boldsymbol{x}') \quad \text{for } x \in V, \tag{2}$$

Journal of Physics: Conference Series **765** (2016) 012006 doi:10.1088/1742-6596/765/1/012006

where $\sigma_{ij}^{(c)}(\boldsymbol{x}')$ is the classical, macroscopic stress tensor at point $\boldsymbol{x}'$, and is equal to $C_{ijkl} \in_{kl} (\boldsymbol{x})$ (C_{ijkl} = the elastic modulus tensor of classical isotropic elasticity and $\in_{kl} (\boldsymbol{x})$ = the strain tensor), $\alpha(|\boldsymbol{x}' - \boldsymbol{x}|, \tau)$ is the nonlocal modulus (or the kernel function), which incorporates the nonlocal effect at the reference point $\boldsymbol{x}$ produced by local strain at the source $\boldsymbol{x}'$ into the constitutive equations. Here $|\boldsymbol{x}' - \boldsymbol{x}|$ is the Euclidean distance and the material constant τ is given by $\tau = e_0 a/l$ [9], where e_0 is a constant appropriate to each material, 'a' is an internal characteristic length (lattice parameter, size of grain, granular distance) and 'l' is an external characteristic length (wave length, crack length, size or dimension of sample) of the system. The value of e_0 needs to be determined from experiments or matching dispersion curves of plane waves with those of atomic lattice dynamics. The volume integral is over the region V occupied by the body. Equation (2) involves spatial integrals which represent weighted averages of the contributions of the strain tensors of all points in the body to the stress tensor at the given point. Hence it is difficult mathematically to get the solution of nonlocal elasticity problems. However, Eringen [8] found that the integral constitutive equation (2) can be converted exactly into equivalent differential form, for some kernels, such as

$$(1 - (e_0 a)^2 \nabla^2)\sigma_{ij}(\boldsymbol{x}) = \sigma_{ij}^{(c)}(\boldsymbol{x}'), \tag{3}$$

where ∇^2 denotes the second order spatial gradient (the Laplacian). The characteristic length $e_0 a$ account for the so-called nonlocal effects. For CNT, the value of e_0 has not been determined by experiment. The value of the adjustable parameter e_0 has been taken as 0.39 by Eringen [8], whereas Sudak has proposed its value in the order of hundreds [5]. Further, the value of parameter 'a' can be chosen to be the distance between $\mathrm{C-C}$ bonds which is equal to 0.142 nm [5]. For carbon nanotubes in one-dimensional case, equation (3) becomes [9-10]

$$\left[1 - (e_0 a)^2 \frac{d^2}{dx^2}\right] \sigma(x) = E\,\epsilon(x). \tag{4}$$

3. General method of determining nonlocal equations of bending of carbon nanotubes

Consider an infinitesimal element PQRS of length dx of a beam under uniformly distribution load p. Let V and $V + dV$ be the shear forces at the left and the right faces of the element during its downward displacement. And the corresponding internal resisting moment at the left face due to the left portion of the beam is M in the clockwise direction and at the right portion of the beam is $M + dM$ in the counter clockwise direction. The free body diagram of the beam element is shown in figure 1.

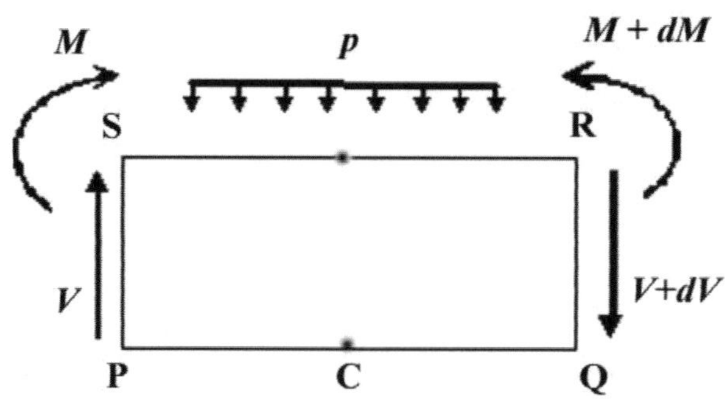

Figure1. An infinitesimal element under uniformly distributed load.

The element must be equilibrium under the action of these forces and couples. The equilibrium equation for the vertical force component is

$$V - pdx - (V + dV) = 0\,.$$

This gives

$$p(x) = -\frac{dV(x)}{dx}. \quad (5)$$

Taking moments about C, the moment equilibrium equation is written as

$$M + V \times \left(\frac{1}{2}dx\right) + (V + dV) \times \left(\frac{1}{2}dx\right) - (M + dM) = 0.$$

Neglecting the product of the infinitesimally small quantities $dV \times dx$, we obtain

$$V(x) = \frac{dM(x)}{dx}. \quad (6)$$

Substituting this into equation (5), we obtain

$$p(x) = -\frac{d^2M(x)}{dx^2}. \quad (7)$$

From definitions, the bending moment M and the axial strain $\epsilon(x)$ for large deflection of the CNT are given by [11]

$$M(x) = \int_A \sigma(x)y\, dA \text{ and } \epsilon(x) = -yK(x), \quad (8)$$

where y is the coordinate measured from the mid-plane in the height direction of the beam, A the area of the cross-section of the beam. The curvature $K(x)$ is related to the vertical deflection $\omega(x)$ of CNT at x by [12]

$$K(x) = \frac{\omega''(x)}{\{1+[\omega'(x)]^2\}^{3/2}}, \quad (9)$$

where (') denotes differentiation with respect to x. Now the Euler-Bernoulli relation between strain and curvature becomes

$$\epsilon(x) = -\frac{y\omega''(x)}{\{1+[\omega'(x)]^2\}^{3/2}}. \quad (10)$$

Combining equations (4) and (10), we have

$$\left[1 - (e_0a)^2\frac{d^2}{dx^2}\right]\sigma(x) = -\frac{Ey\omega''(x)}{\{1+[\omega'(x)]^2\}^{3/2}}. \quad (11)$$

Multiplying both sides of equation (11) by y and integrating over A, we obtain

$$\int_A \sigma(x)ydA - \int_A (e_0a)^2\frac{d^2\sigma(x)}{dx^2}ydA = \int_A -\frac{Ey\omega''(x)}{\{1+[\omega'(x)]^2\}^{3/2}}ydA.$$

By using the definition of bending moment the above equation becomes

$$M(x) - (e_0a)^2\frac{d^2M(x)}{dx^2} + EI\frac{\omega''(x)}{\{1+[\omega'(x)]^2\}^{3/2}} = 0,$$

where $I = \int_A y^2dA$ is the moment of inertia and EI is the bending rigidity of the beam structure. Using equation (7), the expression for the nonlocal bending moment is given by

$$M(x) = -(e_0 a)^2 p(x) - EI \frac{\omega''(x)}{\{1+[\omega'(x)]^2\}^{3/2}} \ . \tag{12}$$

Using equation (6) the nonlocal shear force $V(x)$ can be expressed as

$$V(x) = -(e_0 a)^2 \frac{\mathrm{d}p(x)}{\mathrm{d}x} - EI \frac{\mathrm{d}}{\mathrm{d}x}\left[\frac{\omega''(x)}{\{1+[\omega'(x)]^2\}^{3/2}}\right] . \tag{13}$$

From equation (5), we have

$$p(x) = -\frac{\mathrm{d}V(x)}{\mathrm{d}x} = (e_0 a)^2 \frac{\mathrm{d}^2 p}{\mathrm{d}x^2} + EI \frac{\mathrm{d}^2}{\mathrm{d}x^2}\left[\frac{\omega''(x)}{\{1+[\omega'(x)]^2\}^{3/2}}\right].$$

Or,

$$EI \frac{\mathrm{d}^2}{\mathrm{d}x^2}\left[\frac{\omega''(x)}{\{1+[\omega'(x)]^2\}^{3/2}}\right] - \left[1-(e_0 a)^2 \frac{\mathrm{d}^2}{\mathrm{d}x^2}\right] p(x) = 0. \tag{14}$$

This is the Euler-Bernoulli beam equation for nonlocal beam model for CNTs subjected to a distributed load. When $e_0 = 0$, the above equation reduces to

$$EI \frac{\mathrm{d}^2}{\mathrm{d}x^2}\left[\frac{\omega''(x)}{\{1+[\omega'(x)]^2\}^{3/2}}\right] - p(x) = 0, \tag{15}$$

which is the local Euler-Bernoulli beam equation for large deflection. Again, for small deflection the slope $\omega'(x)$ is small, we can neglect $[\omega'(x)]^2$ in comparison with 1 in equation (14) and obtain the following equation for nonlocal case

$$EI \ \frac{\mathrm{d}^4 w(x)}{\mathrm{d}x^4} - \left[1-(e_0 a)^2 \frac{\mathrm{d}^2}{\mathrm{d}x^2}\right] p(x) = 0. \tag{16}$$

And the local Euler-Bernoulli beam equation for small deflection becomes

$$EI \ \frac{\mathrm{d}^4 w(x)}{\mathrm{d}x^4} - p(x) = 0. \tag{17}$$

As a simple example, we study small-scale effects on the response of a cantilevered CNT of length L clamped at $x = 0$ and free at the other used as an actuator. When a concentrated force $P(x)$ is applied at x on the beam the governing equation (14) gives

$$EI \frac{\mathrm{d}^2}{\mathrm{d}x^2}\left[\frac{\omega''(x)}{\{1+[\omega'(x)]^2\}^{3/2}}\right] - \left[1-(e_0 a)^2 \frac{\mathrm{d}^2}{\mathrm{d}x^2}\right] P(x) = 0. \tag{18}$$

The load applied at $x = l \ (l < L)$ can be written as $P(x) = P\delta(x - l)$, where δ is the Dirac delta function. Then equation (18) becomes

$$EI \frac{\mathrm{d}^2}{\mathrm{d}x^2}\left[\frac{\omega''(x)}{\{1+[\omega'(x)]^2\}^{3/2}}\right] = P\left[1-(e_0 a)^2 \frac{\mathrm{d}^2}{\mathrm{d}x^2}\right] \delta(x - l).$$

Integrating the above equation we have

$$EI \frac{\mathrm{d}}{\mathrm{d}x}\left[\frac{\omega''(x)}{\{1+[\omega'(x)]^2\}^{3/2}}\right] = P[H(x - l) - (e_0 a)^2 \delta'(x - l)] + C_1 \ , \tag{19}$$

where the Heaviside function $H(x-l)=\int \delta(x-l)dx$ and C_1 is a constant of integration. Integrating equation (19) again we have

$$EI\left[\frac{\omega''(x)}{\{1+[\omega'(x)]^2\}^{3/2}}\right]=P[(x-l)H(x-l)-(e_0a)^2\delta(x-l)]+C_1x+C_2\,. \tag{20}$$

This is a non-linear differential equation governing ω(x). Putting $u(x)=\omega'(x)$, the above equation becomes

$$\frac{du(x)}{\{1+[u(x)]^2\}^{3/2}}=\frac{1}{EI}\{P[(x-l)H(x-l)-(e_0a)^2\delta(x-l)]+C_1x+C_2\}dx\,.$$

On further integration we obtain

$$\frac{u(x)}{\{1+[u(x)]^2\}^{1/2}}=\frac{1}{EI}\left\{P\left[\frac{(x-l)^2}{2}H(x-l)-(e_0a)^2H(x-l)\right]+C_1\frac{x^2}{2}+C_2x+C_3\right\}.$$

where C_3 is a constant of integration. Replacing $u(x)$ by $\omega'(x)$ we have

$$\frac{\omega'(x)}{\{1+[\omega'(x)]^2\}^{1/2}}=\frac{1}{EI}\left\{P\left[\frac{(x-l)^2}{2}H(x-l)-(e_0a)^2H(x-l)\right]+C_1\frac{x^2}{2}+C_2x+C_3\right\}\equiv f(x)\,. \tag{21}$$

On squaring equation (21) becomes

$$\frac{[\omega'(x)]^2}{1+[\omega'(x)]^2}=[f(x)]^2.$$

Simplifying the above equation we obtain

$$\omega(x)=\int\frac{f(x)}{\{1-[f(x)]^2\}^{1/2}}dx+C_4\,, \tag{22}$$

where C_4 is the constant of integration. All the constants of integrations C_i $(i=1,2,3,4)$ can be determined from the boundary conditions. For the cantilevered beam, the boundary conditions are

$$\omega(x)|_{x=0}=\omega'(x)|_{x=0}=0 \text{ and } M(x)|_{x=L}=V(x)|_{x=L}=0\,. \tag{23}$$

The first boundary conditions give $C_3=C_4=0$ and the second boundary conditions give $C_1=-P$ and $C_2=PL$. Thus equation (23) becomes

$$\omega(x)=\int\frac{g(x)}{\{1-[g(x)]^2\}^{1/2}}dx$$

where

$$g(x)=\frac{P}{EI}\left\{\left[\frac{(x-l)^2}{2}H(x-l)-(e_0a)^2H(x-l)\right]-\frac{x^2}{2}+Lx\right\}. \tag{24}$$

This is the expression of total deflection in nonlocal models for a cantilevered CNT. Equation (24) shows that the small scale term vanishes in the domain $x<l$, but it will affect when $x>l$.

3.1. Expression for upper bound on the load

The value of total deflection of the bent CNT in equation (24) should be real. This puts a restriction on the value of the denominator. This gives

$$[g(x)]^2<1.$$

Taking the square root we obtain

$$g(x) = \frac{P}{EI}\left\{\left[\frac{(x-l)^2}{2}H(x-l)-(e_0a)^2H(x-l)\right] - \frac{x^2}{2} + Lx\right\} < 1. \quad (25)$$

For the load at the tip of the free end of the CNT, $x = L$. This gives

$$P_{max} < \frac{2EI}{[L^2-(e_0a)^2]}. \quad (26)$$

This is the expression of upper bound on the load in nonlocal models that can be applied at the tip of the free end of a cantilevered CNT which can be regarded as the maximum load needed for breaking the tube. It is evident that the small-scale term has an effect on the value of the maximum load for CNTs having small length (less than 100 nm). When $e_0 = 0$, the above equation reduces to

$$P_{max} < \frac{2EI}{L^2}. \quad (27)$$

This gives the maximum load needed for breaking the CNT which depends on the rigidity EI and on the length L of the CNT as is the case in classical theory.

4. Conclusions

By using the nonlocal Euler-Bernoulli beam theory, we have derived analytical expressions for large deflection for a cantilevered CNT. It is found that there is an upper bound on the load that can be applied at the tip of the free end of a cantilevered CNT which can be regarded as the maximum load needed for breaking the tube. Our results can be used for studying nanoelectromechanical cantilevered systems of small length scale.

Aknowledgement

The author thanks B. R. K. L. Lyngdoh Mawphlang for assistance.

References

[1] Wong E W, Sheehan P E and Lieber C M 1997 *Science* **277** 1971
[2] Poncharal P, WangZ, Ugarte D and de Heer W A 1999 *Science* **283** 1513
[3] Falvo M R, Clary G J, Taylor R M, Chi V, Brooks F P, Washburn S and Superfine R 1997 *Nature* **389** 582
[4] Patra P K and Kumar S 2013 *J. Comput. Theor. Nanosci.* **10** 2732
[5] Sudak L J 2003 *J. Appl. Phys.* **94** 7281
[6] Govindjee S and Sackman J L 1999 *Solid State Commun.* **110** 227
[7] Eringen A C, 1972 *Int. J. Eng. Sci.* **10** 1
[8] Eringen A C 1983 *J. Appl. Phys.* **54** 4703
[9] Peddieson J, Buchanan G R and McNitt R P 2003 *Int. J. Eng. Sci.* **41** 305
[10] Wang Q and Shindo Y,2006 *J. Mech. Mater. Struct.* **1** 663
[11] Weaver Jr W, Timoshenko S P and Young D H 1990 *Vibration Problems in Engineering* (New York: Wiley)
[12] Elmore W C and Heald M A 1969 *Physics of Waves* (New York: Dover)

Effect of ion irradiation on nanoscale TiS_2 systems with suppressed Titania phase

Saurabh J Hazarika[1], Dambarudhar Mohanta[1], A Tripathi[2], D. Kanjilal[2]

[1]Dept. Of Physics, Nanoscience and Soft Matter Laboratory Tezpur university, Napaam-784028, India
[2]Inter University Accelerator Centre, P.O. Box 10502, Aruna Asaf Ali Marg, New Delhi 110067, India

best@tezu.ernet.in

Abstract. Titanium disulfide (TiS_2), being an important of the transition metal dichalcogenide, (TMDC) family, has drawn numerous interest owing to exhibition of tunable band gap as well as high carrier mobility. In this work, we highlight preparation of TiS_2 nanopowder with minimal TiO_2 content and also demonstrate modified properties upon swift heavy ion irradiation on TiS_2 nanoparticles dispersed PVA films. Different properties of the irradiated samples have been characterized through diffraction, microscopic and spectroscopic techniques. As a result of irradiation, due to agglomeration of particles, the grain size is found to increase. We could also observe a red shift after irradiation with increasing fluence, leading to easy flow of electron from valence to conduction band, which shows that conduction of electrons is more in case of irradiated films compared to the pristine one and thus there may be a possibility of using the irradiated samples in various optoelectronic devices.

1. Introduction

In the recent years, 2D materials like graphene have received a great deal of attention among the researchers due to their wide range of electronic and optical properties [1]. However, due to the lack of band gaps, the electronic devices made from graphene cannot be effectively switched on and off and have low turn on and off time ratios [1]. There is a need to overcome these limitations using other layered structures. This leads to the exploitation of other 2D materials, such as, transition metal dichalcogenides (TMDC) which possess sizable band gaps and thus can be promising for electronic and optoelectronic components [1]. Moreover the band gap of these materials can be tuned by tuning their dimensions and thus making them versatile for their use in different applications [2]. TMDCs have generalised formula MX_2, where M is a transition metal element (group IV or group V), and X is a chalcogen [3]. Among the TMDCs, titanium disulphide (TiS_2) has been identified as an active cathode in lithium-based rechargeable batteries [4]. These materials are layered structures of the form S-Ti-S unit, the sulphur atoms in two hexagonal planes separated by a plane of Ti atoms. The Ti-S bonds are bonded covalently whereas the adjacent S-Ti-S layers are bonded through the weak Vander Waals forces. The properties like high conductivity as well as high lithium diffusion rates have made TiS_2 to perform as a stable and high electrode material with long life cycle [5]. As, it may be noted, that TiS_2 can also serve as an efficient catalysis for hydrogenation, and as a hydrogen storage material

[6]. The study of TMDCs including TiS_2 is gaining momentum owing to its versatility and being regarded as an alternative to the zero gap graphene based material.

All materials possess structural defects and due to these defects, materials possess several interesting properties [7]. Tuning of defects within TMDC devices serves as an alternative method to vary electronic and optoelectronic properties [8]. These may be tuned through irradiation effects. Ion irradiation, apart from defect formation and alteration, also cause several other significant changes, which include track formation in material, implantation etc [9]. Bombardment with ions brings in changes in the electronic, optoelectronic properties, apart from structural morphology. Irradiation with charged particle beams allows precise control of defect generation by altering beam conditions and exposure dose [8]. Kim et al. have shown the use of a high energy proton beam to introduce trap states in the back gate dielectric of a MoS_2 thin-film transistor [10]. Tongay et al. have used α-particle irradiation to generate vacancies in TMDCs, which give rise to new emission peaks with enhanced photoluminescence intensity [11]. Fox et al. have demonstrated the use of a focused helium-ion beam to pattern MoS_2 as well as preferentially sputter sulfur atoms [12]. The local tuning of opto-electronic properties of mono and few-layer TMDCs can provide an excellent opportunity to realize sharp homo-junctions similar to conventional p-n, p-i-n, or p-n-p junctions, which are critical to many device architecture [8].

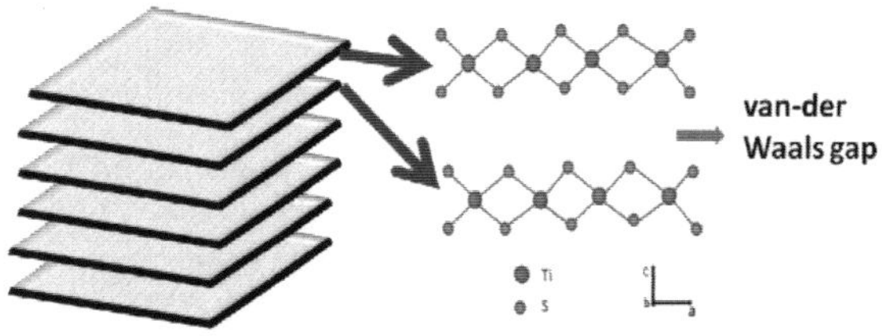

Figure 1. Schematic diagram of TiS_2

Among all other TMDCs, TiS_2 is very less stable as it gets oxidized easily and forms TiO2. Therefore the preparation of TiS_2 needs no exposure of oxygen. The synthesis of TiS_2 at room temperature has not been made till now. In this work, we made several attempts to synthesize TiS_2 nanoscale system with reduced titania phase. The synthesized nanoparticles were first dispersed in a matrix and then bombarded with the 80-MeV nitrogen ion at different fluencies. The modified structural, morphological and optical properties are discussed.

2. Experimental: Materials And Methods

2.1. Synthesis of TiS_2 nanoparticles and sheets

As for synthesis, TiCl4 (LobaChemie, 99.5% pure) as the titanium source and Na_2S as the sulphur source while tetrahydrofurane (THF) (THF, Rankem, 99.7%pure) was taken as the solvent knowing that a polar solvent is free from oxygen and water [13]. The as-received, Na_2S is first dried at 150□C in vacuum for some hours to remove the surface adsorbed water and making Na_2S perfectly anhydrous. Then 1.2 g of the anhydrous Na_2S is taken in 10 ml of THF and is subjected to magnetic stirring at a temperature of 50° C. After about 30 minutes, 1ml of TiCl4 is added to the solution under stirring and left undisturbed for 12h at room temperature. The precursor is then centrifuged (~5000 rpm) and washed with methanol and THF for several times and dried in vacuum. The powder obtained is then labeled as S_1. While preparing the sample, care was taken to avoid direct exposure to air oxygen in it. The synthesis steps are illustrated in the block diagram shown in figure 2.

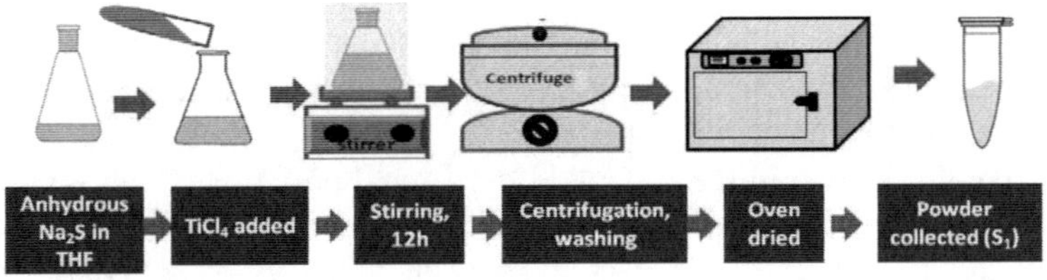

Figure 2. Block diagram of the synthesis steps of sample S_1

2.2. Irradiation of the synthesised nanoparticles

Polyvinyl Alcohol (PVA) was selected to disperse the TiS_2 nanoparticles so as to protect against particle agglomeration and environmental attack [9]. For the irradiation purpose, TiS_2 nanoparticles were dispersed in PVA films (50% volume dispersion) and then casted on laboratory glass slides. The sample so prepared is labeled as sample S_2. The samples were irradiated in the Material Science chamber under a high vacuum (pressure of 10^{-6} mbar) condition and using 80 MeV-N^{4+} ion beams (with a beam current of 1 pnA, particle-nanoampere), available at the 15UD tandem pelletron accelerator of Inter University Accelerator Centre, New Delhi. The ion beam fluence was measured by integrating the ion charge on the sample ladder, which was insulated from the chamber. The ion fluence was varied as 5×10^{11} ions/cm^2, 5×10^{12} ions/cm^2 and 5×10^{13} ions/cm^2 and accordingly the samples were named as S_3, S_4 and S_5. The schematic diagram of the ion beam falling on the films is shown in figure 3.

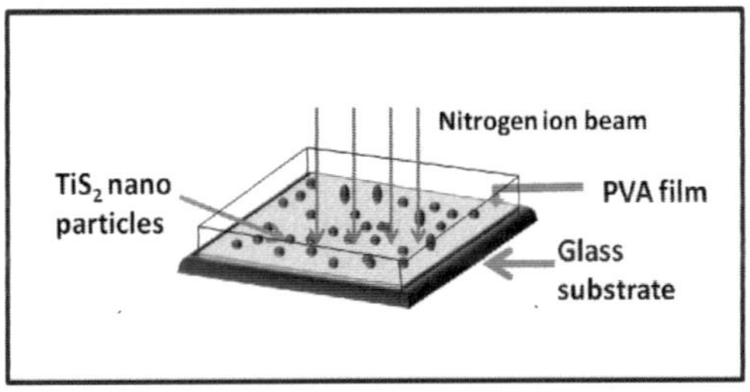

Figure 3. Schematic diagram of nitrogen ion beam incident on the film

3. Characterisation Techniques Employed

To exploit information as regards structural and crystallographic phase, x-ray diffraction (XRD) study was carried out by employing a MiniFlex Rigaku X-ray diffractometer equipped with a CuK_α source (λ=1.543 Å). Morphological analysis was carried out both by scanning electron microscopy (JEOL, 6390LV), Field emission scanning electron microscopy (MIRA II LMH from TESCAN), with a resolution of 1.5 nm at 30 kV was employed to study the irradiated films at a better magnification. It has a secondary electron (SE) and a backscattered electron (BSE) detector for imaging. An energy dispersive X ray detector has also been installed in this system for elemental analysis and transmission electron microscopy (FEI, Tecnai). The elemental analysis was performed through the energy dispersive x-ray spectroscope (EDX) (INCA PentaFET3 with 133 keV resolution from OXFORD) attached to the SEM machine To study the optical properties, we have employed the UV–Visible absorption spectroscopy (UV 2450, Shimadzu Corporation).

4. Results And Discussions

4.1. Structural and elemental analyses

The diffractogram of the pristine samples (S_1, S_2) and the irradiated samples(S_3,S_4,S_5) at different fluences are shown in figure 4 In sample S_1, the diffraction peaks positioned at 2θ ~15.30°, 34.31°, 44.34°, 53.51° and 65.35° characterize hexagonal phase of TiS_2 (figure 4(a)) (JCPDS card no. 74-1141) [6]. Most of the crystallites are believed to have preferential orientation along the (011) plane.

A suppressed NaCl peak at ~31.80° is also seen. Moreover, a small peak at 48.20° corresponds to the anatase phase of TiO_2 [14]. On the other hand for the TiS_2/PVA system (S_2), the peaks for TiS_2 are observed at 2θ ~15.35°, 34.12°, 44.15° and 65.25^0 along with the characteristic crystalline peak of PVA is observable at 2θ ~20.30° [15] (figure 4(b)). With irradiation, all these peaks are found to shift towards a higher angle and decrease in the full width half maxima FWHM value with increasing the ion fluence (figure. 4(c) and 4(d)). But, no additional peak has been observed as a result of irradiation. The peaks shifting towards higher angle suggests an increase in the average crystallite size as well as decrease in microstrain in the irradiated films [16]. An observable decrease in the FWHM value reveals the increase in crystallinity which can also be witnessed by the increase of the peak intensity as we go on incresing the ion fluence. The increase in the grain size can be explained by the fact that with the impact of high energetic ion on to the PVA films the polymer gets destroyed, allowing TiS_2 nanoparticles free from encapsulation (figure. 4 (b)). These irradiation induced free particles can agglomerate and make a bigger particle and thus increasing the grain size [17].

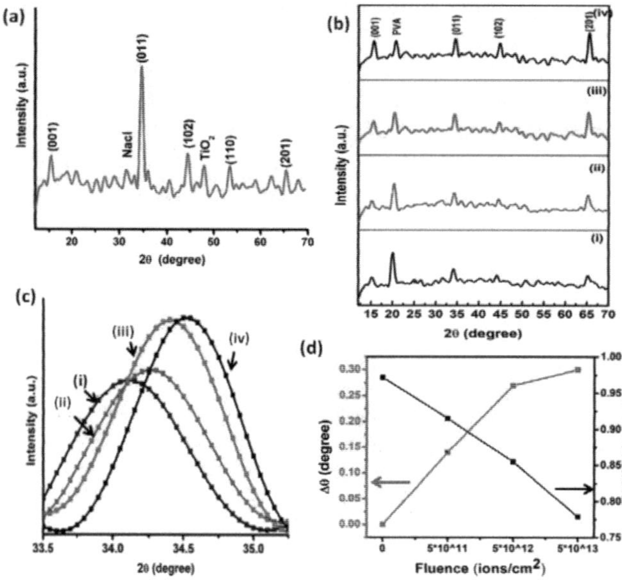

Figure 4. (a) XRD pattern of the sample S_1, (b) XRD patterns of the samples S_2 (i), S_3 (ii), S_4 (iii) and S_5 (iv), (c) shifting of the (011) peak, (d) plot showing variation of peak shift and FWHM with ion fluences

Table 1: Physical parameters obtained through XRD analysis

Ion fluence (ions/cm^2)	Peak position, 2θ (degree)	FWHM, β (degree)	Microstrain (ε) ($\times 10^{-3}$)	Lattice parameters (Å)	Crystallite size, d (Å)
0	34.12	0.972	4.05	a = 3.48, c = 5.71	85.18
5×10^{11}	34.26	0.915	3.81	a = 3.46, c = 5.67	90.43
5×10^{12}	34.39	0.854	3.55	a = 3.45, c = 5.65	96.92
5×10^{13}	34.51	0.772	3.24	a = 3.43, c = 5.63	106.33

Using single line fitting and Debye-Scherrer formula ($d = 0.9\ \lambda/\ \beta\ \cos\theta$, where, λ is the wavelength of the x-ray source, β is the full width at half maxima (FWHM) in radians and θ is the Bragg's angle), the average crystallite size, d is estimated to be ~85.18 Å for the sample S_2, whereas for the irradiated samples, the crystallite is found to increase with increase in ion fluence. The average crystallite size was found to be 90.43 Å, 96.92 Å and 106.33 Å for S_3, S_4 and S_5 respectively. The lattice parameters for each of the sample have been calculated and are given in the Table 1. We have also calculated the microstrain ($\boldsymbol{\varepsilon}$) present in the films by the relation, $\boldsymbol{\varepsilon} = \beta \cos \theta/4$. From the calculation, the microstrain was found to decrease with increase in ion fluence. The value of FWHM, microstrain and the average crystallite size for the unirradiated and the irradiated samples are summarized in the table 1.

4.2. Morphological analyses through electron microscopy imaging

Figure 5 shows the TEM and SEM micrograph of a thin sliced specimen of the sample S_1. The development of TiS_2 nanosheets along with foldings is apparently realized from the micrograph shown in figure 5(a). A magnified image is also shown in figure 5(b). It highlights multilayer, bilayer sheets. The TEM image figure [5(d)] shows that the interlayer separation of TiS_2 ~0.567 nm, which is in good agreement with the literature value for the (001) of TiS_2 [18]. The selected area electron diffraction pattern (SAED) is shown in figure 5(c).

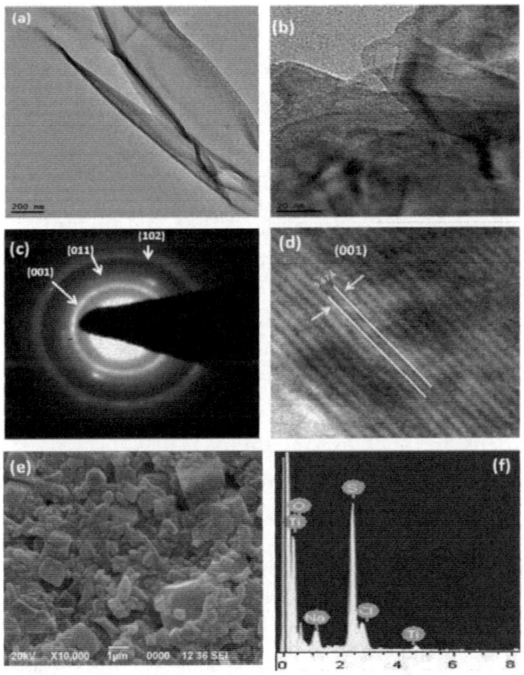

Figure 5. TEM images of sample (a) S_1 (b) Higher magnification image of S_1, (c) SAED pattern of the sample S_1, (d) interlayer spacing and (e,f) SEM images of the sample S_1

The diffused, yet distinguishable diffraction ring patterns suggest polycrystalline nature of the product along with the small amount of single crystalline content. The rings represent the (001), (011) and (012) planes of the synthesized TiS_2 sample.

The SEM micrograph of the sample S_1 is shown in figure 5(e). One can visualize micron-to submicron size distribution of TiS_2 particles (figure 5 (e)). Typically, the average size of the unclustered particles ranges between 50 nm- 200 nm. At a higher magnification, some of the particles appeared to be in agglomerated form, with a diverse morphology, and having dimension < 2 μm. The agglomerated particles tend to form hexagonal prism-like conformation owing to the hexagonal phase of TiS_2 [3, 6]. The presence of Ti and S in the sample S_1 can be witnessed from the EDX micrograph shown in figure 5(f). The FESEM micrographs of the samples S_2 and S_4 are shown in figure 6 (a, b, c, d).

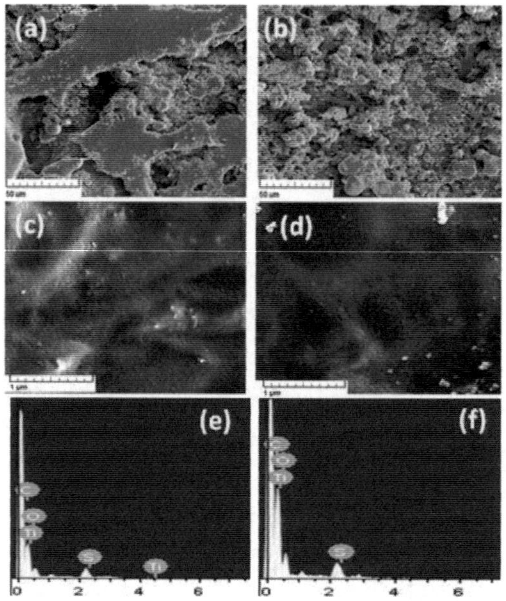

Fig. 6: FESEM images of the samples (a,b) S_2, (c,d) S_4 and EDX micrographs of the samples (e) S_2, (f) S_4

It is found that with irradiation, the PVA film got damaged Fig. 6(a, c) and the size of the TiS_2 particles are enhanced Fig. 6(b, d) in the sample S_4 compared to S_2. The enhancement of the particle size is due the agglomeration of the TiS_2 nanoparticles.

The EDX spectrum of the pristine sample (S_1) reveals a proportionately high intense S peak along with Ti peak and suppressed Na and Cl peaks (Fig. 6(e)). Quantitatively, Ti to S atomic percentage ratio is nearly 1:2, as desired for the development of TiS_2.Whereas, the evidence of presence of C, O, Ti and S can be assured from the EDX spectrum of the irradiated sample (Fig. 6(f).

4.3. UV- Visible optical absorption spectroscopy analysis

The UV Vis spectra of the pristine and irradiated samples are shown in figure 7.

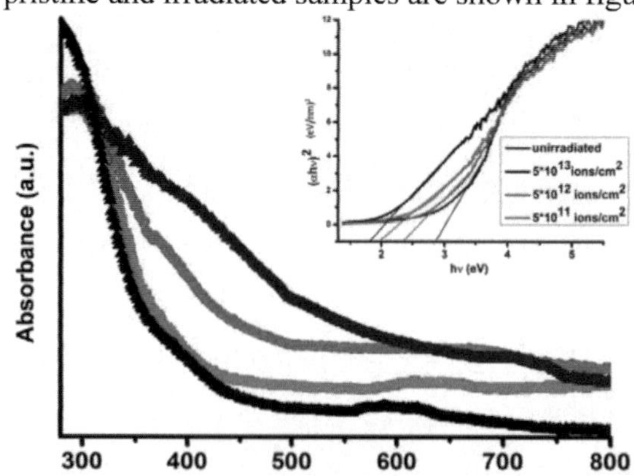

Figure 7: FESEM images of the samples (a,b) S_2, (c,d) S_4 and EDX micrographs of the samples (e) S_2, (f) S_4

It is found that the pristine sample shows an absorbtion peak at 596 nm which is ascribed due **to** $\mathbf{M_2^- - M_1^+}$ of TiS_2 transition [19]. With an increase in the fluence, a red shift in the absorbtion edge is occurred which leads to the decrease in the bandgap of the TiS_2 film. The optical band gap E_g can be calculated using the Tauc's relation for direct and allowed transition [20]

$$\alpha h\upsilon = A\,(h\upsilon - E_g)^{1/2}$$

where, $h\upsilon$ is the energy of the photon and A is the system dependent parameter. From the Tauc's plot (shown inset) the bandgap of the unirradiated sample (S_2) was found to be 2.85 eV and the value was observed to be decrease with increase in the ion fluence. The value of the bandgaps are found to be 2.34 eV, 1.97 eV and 1.79 eV for the samples S_3, S_4 and S_5 respectively. The decrease in the band gap of the irradiated sample may be due to the increase in the crystallite size of the samples [21]. It can be easily understood that with decrease in the band gap the transfer of electrons from the valence band to conduction band can easily take place. Thus the conduction of electrons will be more in the irradiated films than the pristine one and thus there may be a possibility of using the irradiated samples in various optoelectronic devices [22].

5. Conclusions

We have synthesized TiS_2 with reduced titania phase by a simple chemical route and characterized the samples with different characterization techniques. The obtained TiS_2 particles are dispersed on PVA and casted on a laboratory glass slide. The resultant TiS_2/PVA films were irradiated using 80 MeV N^{4+} ions at different fluences. The irradiated films were characterized using XRD, EDX, SEM and UV Vis spectroscopy. Due to irradiation, the agglomeration of the TiS_2 particles took place and the grain size was found to increase as evident from XRD and SEM analysis. A gradual decrease in the band gap with increase in ion fluence was observed from the Tauc's plot which is the result of the increase in the grain size of the TiS_2 particles.

Acknowledgments

The authors acknowledge IUAC, New Delhi for the financial support (UFR-56322/2014) and thank them for providing good quality nitrogen beam and FESEM and EDX facilities. The authors thank SAIC, TU, for extending TEM imaging facility. We also thank colleagues for valuable discussion and timely assistance.

References

[1] Wang Q H, K-Zadeh K, Jonathan A K, Coleman N and Strano M S, "Electronics and optoelectronics of two-dimensional transition metal dichalcogenides", *Nature Nanotechnology*, vol **7**, pp. 699-712, November 2012.

[2] Sajanlal P R, Sreeprasad T S, Samal A K and Pradeep T, "Anisotropic nanomaterials: structure, growth, assembly, and functions", *Nano Rev.*, vol **2**, February 2011.

[3] Bhattacharyya S and Singh A K, "Semiconductor-metal transition in semiconducting bilayer sheets of transition-metal dichalcogenides", *Phys. Rev. B.* vol. **86**, pp. 075454(1)- 075454(7), August 2012.

[4] Sriram M A and Kumta P N, "The thio-sol-gel synthesis of titanium disulfide and niobium disulfide . Part 1.–Materials chemistry", *J. Mater. Chem.* Vol **8**, pp. 2441-2451, August 1998.

[5] Winter R and Heitjans P, "Li^+ Diffusion and Its Structural Basis in the Nanocrystalline and Amorphous Forms of Two-Dimensionally Ion-Conducting Li_xTiS_2", J. Phys. Chem. B., vol. 105, pp. 6108-6115, August 2001.

[6] Tan M, Wang Z, Peng J, and Jin X, "Facile Synthesis of Large and Thin TiS_2 Sheets via a Gas/Molten Salt Interface Reaction", *J. Am. Ceram. Soc.*, vol. **98**, pp. 1423-1428, 2015.

[7] Lin Z, Carvalho B R, Kahn E, Lv R, Rao R, Terrones H, Pimenta M A and Terrones M, "Defect

engineering of two-dimensional transition metal dichalcogenides", *2D Mater.*, vol. **3**, pp. 022002(1)- (21), 2016.

[8] Stanford M G, Pudasaini P R, Belianinov A, Cross N, J. H. Noh, Koehler M R, D. Mandrus G, Duscher G, Rondinone A J, Ivanov I N, T. Ward Z and Rack P D, "Focused helium-ion beam irradiation effects on electrical transport properties of few-layer WSe2: enabling nanoscale direct write homo-junctions", *Scientific Reports*, vol. **6**, pp. 272276(1) – (10), June 2016.

[9] Bayan S, Mohanta D, "Effect of 80-MeV nitrogen ion irradiation on ZnO nanoparticles: Mechanism of selective defect related radiative emission features", *Nuclear Instruments and Methods in Physics Research B*, vol. **269** pp.374–379, November 2010.

[10] Kim T Y, K Cho, Park W, Park J, Song Y, Hong S, Hong W K, Lee T, "Irradiation Effects of High-Energy Proton Beams on MoS2 Field Effect Transistors", *ACS Nano*, vol. **8**, pp. 2774–2781, February 2014.

[11] Tongay S, Suh J, Ataca Can, Fan W, Luce A, Kang J S, Liu J, Ko C, Raghunathanan R, Zhou J, Ogletree F, Li J, Grossman J C, Wu J, "Defects activated photoluminescence in two-dimensional semiconductors: interplay between bound, charged, and free excitons", *Sci. Rep.*, vol. **3**, pp. 2657(1)- (5), September 2013.

[12] Fox D S, Zhou Y, Maguire P, O'Neill A, Ó'Coileáin C, Gatensby R, Glushenkov A M, Tao T, Duesberg G S, Shvets I V, Abid M, Abid M, Wu H-C, Chen Y, Coleman, Donegan J F, Zhang H, "Nanopatterning and Electrical Tuning of MoS2 Layer with a Subnanometer Helium Ion Beam", *Nano Lett.*, vol. **15**, pp. 5307–5313, July 2015.

[13] Wold A and Dwight K, "Low Temperature Synthesis of Transition Metal Sulfides", *J. Sol. St Chem.*, Vol. **96**, pp. 53-58, January 1992.

[14] Ramchiary A, Samdarshi S K, "High visible light activity of hydrogenated structure-engineered mixed phase titania photocatalyst", *Chem. Phys. Lett.*, vol. **597**, pp. 63-68, March 2014.

[15] Attia G and Abd El-kader M F H, "Structural, Optical and Thermal Characterization of PVA/2HEC Polyblend Films", *Int. J. Electrochem. Sci.*, vol. **8**, pp. 5672 – 5687, April 2013.

[16] Rana D S, Chaturvedi D K, Quamara J K, "XRD and SEM Investigation of Swift Heavy Ion Irradiated Polyvinylidene Fluoride Thin Films", *Journal of Materials Engineering and Performance*, vol. **20**, pp. 276-282, March 2011.

[17] Nguyen T P, Choi S, Jeon J-K, Kwon K C, Jang H W, and Kim S Y, "Transition Metal Disulfide Nanosheets Synthesized by Facile Sonication Method for the Hydrogen Evolution Reaction", *J. Phys. Chem. C*, vol. **120**, pp. 3929−3935, February 2016.

[18] Wilson J A and Yoffe A D, "Colloidal synthesis of inorganic fullerene nanoparticles and hollow spheres of titanium disulfide", *Adv. Phys.*, vol. **18**, pp. 193- 335, May 1969.

[19] Myron H W, Freeman A J, "Electronic Structure and Optical Properties of Layered Dichalcogenides: TiS_2 and $TiSe_2$", *Phys. Rev. B,* vol. **9**, pp. 481−486, January 1974.

[20] Mir F A, Rehman S, Asokan K, Khan S H, Bhat G M, "Optical, DC and AC electrical investigations of 4-hydroxy coumarin molecule as an organic Schottky diode", *J Mater Sci: Mater Electron*, vol. **25**, pp. 1258–63, January 2014.

[21] Sheena P A, Priyanka K P, Sabu N A, Gannesh S and Varghese T, "Effect of electron beam irradiation on the structure and optical properties of nickel oxide nanocubes", *Bull. Mater. Sci.*, Vol. **38**, pp. 825–830, August 2015.

[22] Kuriakose S, Avasthi D K and Mohapatra S, "Effects of swift heavy ion irradiation on structural, optical and photocatalytic properties of ZnO–CuO nanocomposites prepared by carbothermal evaporation method", *Beilstein J. Nanotechnol.,*vol. **6**, pp. 928–937, April 2015.

Electronic and optical properties of $CuInTe_2$

A. Shankar[1]*, R. K. Thapa[2] and P. K. Mandal[1]

[1]Department of Physics, University of North Bengal, Darjeeling -734013
[2]Condensed Matter Theory Research Group, Physics Department, Mizoram University-796004

*Email: amitshan2009@gmail.com

Abstract. The electronic and optical properties of a ternary chalcopyrite compound $CuInTe_2$ with diamond like structure have been studied. The calculations are carried out using the density functional theory (DFT) based full potential-linearized augmented plane wave (FP-LAPW) method within the framework of GGA and modified Becke Johnson (mBJ) potential approach. The presence of direct energy band gap of 0.8 eV suggests the sample material can be a good material for solar cell application. The study of the optical response of the material against the incident photon energy radiation indicates the material can be an effective candidate for the optoelectronic devices.

1. Introduction

Chalcopyrite compounds form a very large and versatile group of materials formed by some members of the I-III-VI_2 family of the periodic table. These materials are distinguished by their tunable direct band gap semiconducting nature, with optimal range of energy band gap for terrestrial solar conversion [1]. With different combinations of the different members of Group I, III and VI of periodic table the band gaps of these materials vary from infrared to ultraviolet spectral range, which make them promising candidate for solar-cells applications [2,3], light-emitting diodes [4,5], nonlinear optics [6], and optical frequency conversion applications in all solid state based tunable laser systems. With further modification by partial substitution of cations or anions, some of the chalcopyrite candidates possess nonlinear optical properties, which also make them promising candidate for optical device and technology [7]. $CuInTe_2$ is also a candidate compound of ternary chalcopyrite (I-III-VI_2), characterized by their direct band gap semiconducting nature and analogs of the II-VI zinc blende compounds. The energy band gap (1.06 eV) close to the optimum value for solar energy conversion makes $CuInTe_2$ an interesting candidate for material research [8] and suitable for the fabrication of solar cells.

The several experimental and theoretical studies have been performed to understand the physical properties of this ternary Chalcopyrite. $CuInTe_2$ was synthesized by Hahn *et al.* [9] for the first time and after that several researcher conducted the theoretical and experimental calculations to investigate the electronic [10], physical [11,12] and optical [13] properties. Lateron, Jaffe and Zunger [14] used the potential variation mixed basis (PVMB) approach to study the chemical trends of the Cu-based ternary chalcopyrite semiconductors. Ahuja *et al.* [15] have used the full-potential linear muffin-tin orbital (FP-LMTO) method to investigate the optical properties of $CuGaS_2$. The series of compounds $CuXS_2$ (X = Al, Ga, In) and $CuAlX_2$(X = S, Se, Te) including $CuGaS_2$ are also investigated to study their electronic and optical properties [16]. Similarly, $CuGaS_2$ and $AgGaS_2$ were studied to explore their structural, electronic and optical properties using the FP-LAPW method by Laksari *et al.* [17].

Following this study, Reshak and Auluck [18] also studied the electronic properties of $CuAlX_2$(X = S, Se, Te) using FP-LAPW method. Recently, Xian-Zhou *et al.* [19] have studied the electronic structure and optical properties of $CuInTe_2$. However their pseudopotential based study suggests the presence of zero energy band gap for this material, which is against its experimental value of 1.06 eV [20]. Similarly, the previous local density approximation (LDA) based study [21] is also not able to predict the precise value of the energy band gap of $CuInTe_2$ (0.18 eV).

The accurate knowledge of the energy band gap and the optical functions of these materials play a significant role to understand their application as mentioned earlier. In this work, we first analyze the electronic properties of $CuInTe_2$ to predict the precise value of energy band gap, based on first-principle calculations. Another concern of this work is to understand the origin of interband transition above the band gap and the optical response of the material for the energy radiation of infrared to ultraviolet range.

2. Computational details

Ternary $CuInTe_2$ crystallizes in tetragonal structure with space group of I-42d (No. 122). The calculations presented here are performed using a semi-relativistic version of the FP-LAPW method implemented in the Wien2k code [22] within the framework of DFT. The exchange-correlation effects are treated by the Perdew– Berke–Ernzerhof generalized gradient approximation (PBE-GGA) [23]. The energy eigen values are converged expanding the wave function in the interstitial region and the plane wave cut-off was set to $R_{MT} \times K_{max} = 7$, Where R_{MT} is the smallest atomic muffin-tin (MT) sphere radius and K_{max} is the maximum value of the wave vector in the plane wave expansion. Non-spherical contributions to the charge density and potential within the MT spheres are considered up to $l_{max} = 6$, while charge density and the potential are expanded as a Fourier series with wave vectors up to G_{max}=12 (a.u.)$^{-1}$. The different MT sphere radii (R_{MT}) used were 2.43 a.u., 2.9 a.u. and 2.43 a.u. for Cu, In and Te, respectively. The self-consistent potentials are calculated on a 17 × 17 × 17 *k*-mesh in the Brillouin zone (BZ) and the convergence criterion is set to 10^{-4} Ry. The experimental lattice constant of a= 6.195 Å and c = 12.418 Å [24] are used to calculate the results presented here.

3. Results and discussions

3.1. Electronic properties

The energy band structure and density of states are calculated and they show similar profile with their previous report. However, the energy band gap of the material is found to be close to 0.2 eV, which is lower than the experimental value with large deviation. Therefore in the present study we have included a new approach of exchange-correlation functional known as mBJ potential [25] in the calculation to find the precise values of the energy band gap. The mBJ calculated density of states (DOS) and energy bands are presented in Figures 1 and 2, respectively. The overall energy band profile matches with their previous report [19].

The energy band structure can be distinguished into three regions, namely core, valence and conduction regions, separated by well defined energy gaps. The lowest energy region, core region is primarily contributed by Te-*s* electronic state as visible from partial DOS plot. The lower valence region is Te-*p* states dominated with small contribution from In-s orbital electrons, however the region just below the Fermi energy level (E = 0 eV) is formed by Cu-*d* states. The region above Fermi energy level (E_F) is dominated by Te. Moreover the Cu-3*d* states are found to split into two sub-bands at -2 eV and -3 eV below Fermi energy level. Here the Cu-3*d* states are found to split due to tetrahedral crystal field effect, which also gives the presence of a narrow band gap at the valence region.

The energy band structure shows the presence of the highly occupied level of the valence band and the lowest level of the unoccupied conduction bands at the gamma point of the crystalline BZ giving

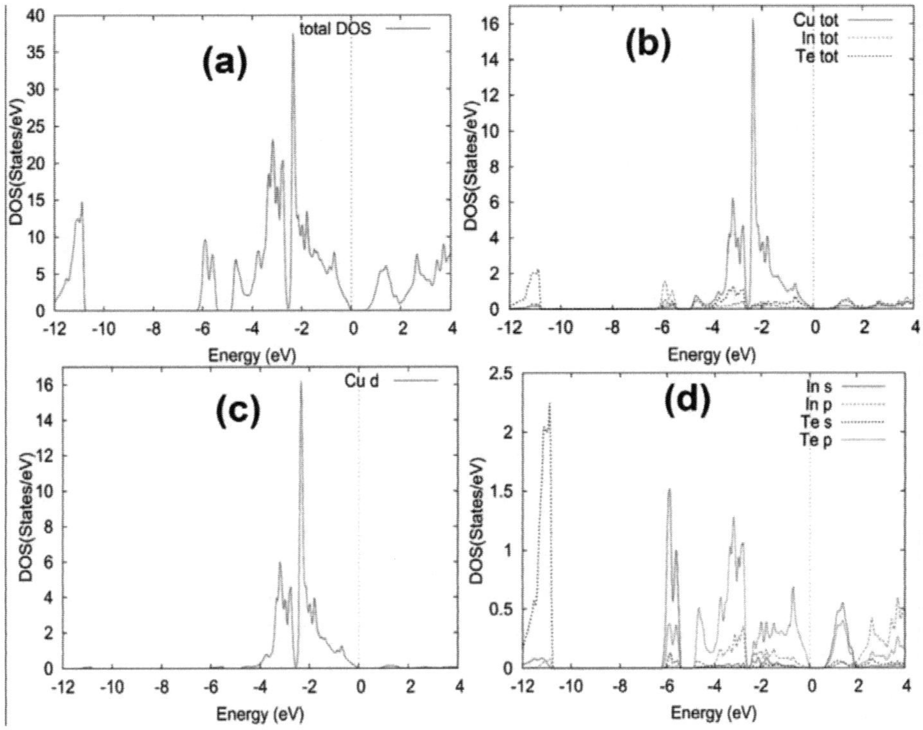

Figure 1. Total and partial density of states.

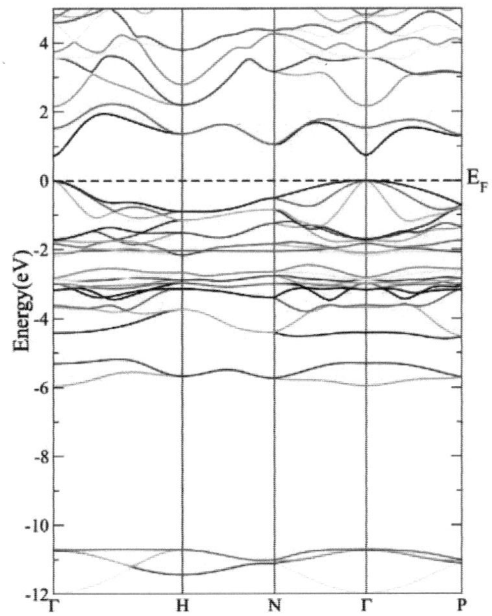

Figure 2. mBJ calculated energy band structure.

an energy band gap of 0.8 eV. The energy band gap of 0.95 eV with values from 0.88 to 1.04 eV have been reported for this compound [26]. However, Xian-Zhou *et al.* [19] have reported the vanishing energy band gap of this material using PBE-GGA as exchange-correlation functional within the framework of ultrasoft pseudopotential method based on DFT. Rashkeev and Lambrecht [21] have estimated the energy band gap of 0.18 eV from the potential-variational mixed basis based method and LDA approach. It is known that LDA underestimate the band gap but here it is seen that LDA is giving better result than previous GGA estimated result [19]. We found that the mBJ calculated gives the better result as compared to previous LDA and GGA estimated values.

3.2. Optical properties

The understanding of the optical response of the material plays a significant role for the optoelectronic application of the material. The dielectric function, which is the most important parameter to describe the optical response of a material, is given by $\varepsilon(\omega) = \varepsilon_1(\omega) + i\varepsilon_2(\omega)$, where, $\varepsilon_1(\omega)$ and $\varepsilon_2(\omega)$ are the real and imaginary parts of the dielectric function. The scissor shift was used as implemented in the computational code used with energy of 1.06 eV to overcome the discrepancy between the experimental and theoretical values of the energy gap. The frequency dependent dielectric function of $CuInTe_2$ is investigated in the energy range 0.0 -12.0 eV, and presented in Figure 3. In the figure (Figure 3) the dielectric functions are plotted with the incident radiation polarized with both perpendicular and parallel to the tetragonal c-axis of the crystal.

The peaks in the spectra correspond to the transition from the valence to the conduction region of the energy bands. From the energy band structure plot one can say that the peaks below 1 eV of energy are due to the optical transitions from the bands just below the E_F to the bands just above the E_F. Similarly, the optical transitions from the middle of the valence band to the conduction band are responsible for the peaks above 1 eV. In the plot one can find that $\varepsilon_1(\omega)$ shows anisotropic behavior in the whole energy range, whereas $\varepsilon_2(\omega)$ shows anisotropic behavior in the inter-mediate energies. The $\varepsilon_2(\omega)$ plot is characterized by the presence of two sharp peaks and shoulder like structure. The first critical point, which corresponds to the threshold for direct optical transitions between the highest occupied states to the lowest unoccupied states, is located at 1.8 eV. The zero frequency limits, $\varepsilon_1(0)$ is the electronic part of dielectric constant, and its value is obtained to be 7.7 and 7.3 for E||C and $E \perp C$, respectively. The obtained value of $\varepsilon_1(0)$ is lower than the previous experimental value of 9.3 [27]. The $\varepsilon_1(\omega)$ spectrum, after certain energy (5.7 eV and 6.1 eV for E||C and $E \perp C$, respectively) tends to fall below unity, which corresponds to the metallic behavior with high value of reflectivity for the incident electromagnetic radiation. So, this compound can be used as a protection from electromagnetic radiations in the specific energy limits.

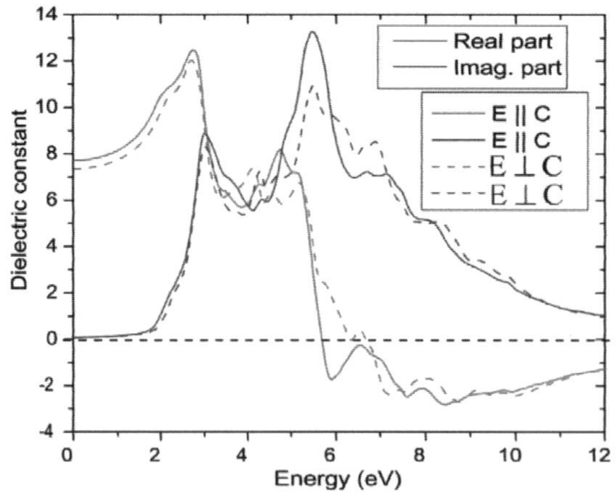

Figure 3. Real and imaginary parts of frequency dependent dielectric constant.

4. Conclusions

The electronic and optical properties of $CuInTe_2$ have been studied by using the first principles calculation. The sample material shows the presence of direct band gap with magnitude of 0.2 eV. The mBJ treatment of the material shows an enhanced value of energy band gap with its value of 0.8 eV, which is close to the experimental result. The investigation of the optical response of the material for the photon energy radiation of infrared to ultraviolet range suggest the metallic nature of the material in the ultraviolet range of radiation with high value of reflectivity.

Acknowledgment

AS would like to acknowledge Dr. D. S. Kothari fellowship, UGC(India) and RKT a SERB(India) project via No.EMR/2015/001407.

References

[1] Shay J L and Wernick J H 1975 Ternary Chalcopyrite Semiconductors: Growth Electronic *Properties, and Applications* (Pergamon Press,Oxford)
[2] Shao L X, Chang K H, Chung T H, Tseng B H, Hwang H L 2003 *J. Phys. Chem.* **64** 1505
[3] Ramanathan K, Hasoon F S, Smith S, Young D L, Contreras M A, Johnson P K, Pudov A O, Sites J R 2003 *J. Phys. Chem. Solid.* **64** 1495
[4] Shay J L, Schiavone L M, Buehier E, Wernick J H 1972 *J. Appl. Phys.* **43** 2805
[5] Wagner S, Shay J L, Tell B, Kasper H M 1973 *Appl. Phys. Lett.* **22** 351
[6] Levine B F 1973 *Phys. Rev.* **B 7** 2600 and references therein
[7] Lin L J , Tabatabaie N, Wernick J H, Hull G W, Meagher B 1988 *J. Elec. Mater.* **17** 321
[8] Mobarak M, Shaban H T 2014 *Mater. Chem. Phys.* **147** 439
[9] Hahn H, Frank G, Klinger W, Merer A and Stroger G 1953 *Z. Anorg. Chem.* **271** 153
[10] Berger L I and Petrov V M 1970 *Neorg. Mater.* **6** 1348
[11] Range K J, Engert G and Weiss A 1969 *Solid. State. Commun.* **7** 1749
[12] Locher P R and Van Stapele R P 1970 *J. Phys. Chem. Solids.* **31** 2643
[13] Koshkin V M, Galchinetski L G, Karas V R, Nesterova T N and Skovskaya I L 1971 *Ukkr. Fiz. Zh.* **16** 1746
[14] Jaffe J E and Zunger A 1984 *Phys. Rev.* **B 29** 1882
[15] Ahuja R, Auluck S, Eriksson O, Wills J M, Johansson B 1998 *Sol. Energy Mater. Sol. Cells* **53** 357.
[16] Zhou H G, Chen H, Chen D, Li Y, Ding K N, Huang X, Zhang Y F 2011 *Acta Phys.-Chim. Sin.* **27** 2805
[17] Laksari S, Chahed A, Abbouni N, Benhelal O, Abbar B 2006 *Comput. Mater. Sci.* **38** 223
[18] Reshak A H, Auluck S 2008 *Solid State Commun.* **145** 571
[19] Xian-Zhou Z, Ke-Sheng S, Zhao-Yong J, Xiao-Fen H 2013 *Comp. Theor. Chem.* **1010** 67
[20] Jaffe J E, Zunger A 1984 *Phys. Rev.* **B 29** 1882
[21] Rashkeev S N, Lambrecht W R L 2001 *Phys. Rev.* **B 63** 165212
[22] Blaha P, Schwarz K, Madsen G K H, Kvasnicka D, Luitz J 2014 *An augmented plane wave + local orbital program for calculating crystal properties* Technische Universitát, Wien
[23] Perdew J P, Wang Y 1992 *Phys. Rev.* **B 46** 12947
[24] Termsaithong P and Tubtimtae A 2015 *Matt. Lett.* **138** 41
[25] Tran F, Blaha P 2009 *Phys. Rev. Lett.* **102** 226401
[26] Kazmerski L L and Juang Y J 1977 *Journal of Vacuum Science & Technology* **14** 769
[27] Davis J G, Bridenbaugh P M, Wagner S 1978 *J. Electron. Mater.* **7** 39

Concentration dependent structural parameters of liquid Al-Fe alloys

C. Lalnuntluanga and Raj Kumar Mishra*

Department of Chemistry, School of Physical Sciences, Mizoram University, Aizawl-796004, India
*Email: rkmishramzu@yahoo.com

Abstract. Square well potential is perturbed over Lebowtiz solution of hard sphere mixtures to determine direct correlation function,$C^{(0)}_{ij}(r)$ in repulsive and attractive regions under Mean Spherical Model Approximation [1]. Obtained direct correlation functions were employed to derive partial structure factors and then total structure factor, S(k) in liquid Al-Fe alloy at different atomic percent of Al. Fourier transform of partial and total structure factors gives partial and total radial distribution functions, g(r) from which partial and total coordination numbers and the partial nearest-neighbor distances were computed.

1. Introduction

Fe-Al alloys are widely used in the production of ductile cast iron, aluminized steel, and corrosion resistant materials, etc. [2, 3]. Due to its unique set of mechanical, magnetic and corrosion properties, the binary Fe-Al system also serve as the basis for the preparation of quasi crystalline and amorphous materials [4].

Since liquid state do not have systematic arrangements of particles, understanding the structure of liquids become all the more complicated and hence their solution and their applications become more important as the theory of symmetry cannot be applied nor the theory of probability in completeness. However, the structural properties of liquids can be explained on the basis of perturbation treatment on hard sphere reference system [5]. In this present work we have performed computations within the framework of square well (SW) model to evaluate partial and total structure factors and their derived associated properties like partial and total radial distribution functions, partial and total coordination numbers of Fe-Al alloys at various atomic fractions of Al.

2. Theory

We perturbed amended square well attractive part over the hard sphere reference system [6] within the Mean Spherical Model Approximation (MSMA) and so the Direct correlation function (DCF) can be written as while charge density and the potential are expanded as a Fourier series with wave vectors up to G_{max}=12 $(a.u.)^{-1}$. The different MT sphere radii (R_{MT}) used were 2.43 a.u., 2.9 a.u. and 2.43 a.u. for Cu, In and Te, respectively. The self-consistent potentials are calculated on a 17 × 17 × 17 *k*-mesh in the Brillouin zone (BZ) and the convergence criterion is set to 10^{-4} Ry. The experimental lattice constant of a= 6.195 Å and c = 12.418 Å [24] are used to calculate the results presented here.

$$C_{ij}(r)=\begin{cases} C^0_{ij}(r) & ; \quad 0<r<\sigma_{ij} \\ -\varepsilon_{ij}/k_B T & ; \quad \sigma_{ij}<r<\lambda_{ij}\sigma_{ij} \\ 0 & ; \quad r>\lambda_{ij}\sigma_{ij} \end{cases} \tag{1}$$

Where, $C^{(0)}_{ij}(r)$ is the DCF for a mixture of hard spheres obtained by Lebowitz for Percus -Yevick equation. Further, σ_{ii}, ε_{ii}, and λ_{ii} are the diameter, the potential energy depth and breadth respectively of the square well potential of the ith species. The mixed parameters are determined by using Lorentz Berthelot rules [7]. The total correlation function $h_{ij}(k)$ is related to the DCF through the generalized Ornstein-Zernikes's equation for a system containing more than one species, which can be given as

$$h_{ij}(k) = C_{ij}(k) + \sum_{1=1,2} \rho_1 C_{i1}(k)\, h_{j1}(k) \tag{2}$$

The Ashcroft-Langreth partial structure factors $S_{11}(k)$, $S_{22}(k)$ and $S_{12}(k)$ can be obtained by taking the Fourier Transformations of $h_{ij}(k)$ given as

$$S_{11}(k) = \left\{1 - \rho_1 C_{11}(k) - \rho_1\rho_2 C^2_{12}(k)/1 - \rho_2 C_{22}(k)\right\}^{-1} \tag{3}$$

$$S_{22}(k) = \frac{[1 - \rho_1 C_{11}(k)]\, S_{11}(k)}{1 - \rho_2 C_{22}(k)} \tag{4}$$

$$S_{12}(k) = \frac{(\rho_1 \rho_2)^{\frac{1}{2}} C_{12}(k) S_{11}(k)}{1 - \rho_2 C_{22}(k)} \tag{5}$$

The total structure factors, S(k) in terms of partial structure factors, $S_{ij}(k)$ and concentration is obtained as follows

$$S(k) = \sum_{i=1}^{2} \sum_{j=1}^{2} (c_i c_j)^{1/2} \frac{f_i(k)\, f_j(k)}{c_1 f_1^2(k) + c_2 f_2^2(k)} S_{ij}(k) \tag{6}$$

Here $f_i(k)$ and $f_j(k)$ are the atomic scattering factors of ith and j[th] species respectively.
We obtain the partial radial distribution functions, $g_{ij}(r)$ and also total radial distribution functions, $g(r)$ by the Fourier inversion of the $S_{ij}(k)$ and S(k) respectively. From $g_{ij}(r)$ we compute the first coordination number ψ_{ij} through the following equation

$$\Psi_{ij} = 4\pi \rho_{ij} \int_0^{r_{min}} g_{ij}(r)\, r^2 dr \tag{7}$$

Here r_{min} is the first minimum of $g_{ij}(r)$.

3. Results and discussion

The computations of the total structure factors were made using the pure parameters for Fe and Al as given in Table 1 at different concentrations of Al in Fe-Al alloys.

Table 1: Potential parameters of Fe-Al alloy

Metals	σ_{ii} (Å)	$\varepsilon_{ii}/K_B(K)$	λ_{ii}	ρ_{ii}
Fe	2.28	425.67	1.76	0.0756
Al	2.40	64.72	1.60	0.0646

It has been found that the computed total structure factors S(k) and the corresponding pair-correlation functions g(r) for liquid Fe-Al alloys at 50 K above the liquidus temperature shows a good agreement with the experimental results [8] at different concentrations of Al in Fe-Al alloys as given in Fig. 1 and Fig. 2. We found that as the atomic percent of Al increases, the peaks occur at the same position but the peak height of the total structure factor increases which is also shown in Fig.1. The peak positions and peak heights of the radial distribution functions of the alloys computed by Fourier transformation of total structure factors increase with increase in the concentration of Al in Fe-Al alloys as given in Table 2. Similar trend was observed experimentally by [8], there is a good agreement between our computed results and the experimental values.

Table 2: Details of structure factor and radial distribution function

% of Al in Fe-Al alloy	Temperature (K)	Theoretical k	Theoretical S(k)	Experimental k	Experimental S(k)	Theoretical r	Theoretical g(r)	Experimental r	Experimental g(r)
70	1493	2.90	2.44	2.99	2.40	2.45	2.95	2.54	3.05
78	1453	2.90	2.48	2.92	2.41	2.52	3.14	2.60	2.77
82	1463	2.90	2.50	2.92	2.49	2.59	3.18	2.63	2.79
86	1433	2.90	2.52	2.92	2.09	2.60	3.23	2.63	2.91

Table3: Partial and total coordination numbers

% of Al in Fe-Al alloy	Ψ_{11}	Ψ_{22}	Ψ_{12}	Ψ_{total}
70	7.961	6.125	0.018	14.104
78	7.960	6.265	0.021	14.246
82	7.945	6.563	0.022	14.531
86	7.949	6.871	0.022	14.842

Table 4: The partial nearest- neighbour distances r (in Å)

% of Al in Fe-Al alloy	$r_{(Fe-Fe)}$ Theoretical	$r_{(Fe-Fe)}$ Experimental	$r_{(Al-Al)}$ Theoretical	$r_{(Al-Al)}$ Experimental	$r_{(Fe-Al)}$ Theoretical	$r_{(Fe-Al)}$ Experimental
70	2.49	2.62	2.33	2.57	2.41	2.53
78	2.62	2.68	2.54	2.59	2.53	2.63
82	2.63	2.71	2.55	2.61	2.53	2 .64
86	2.54	2.68	2.57	2.60	2.54	2.65

From partial radial distribution function we compute the coordination number of the alloys at different concentration of Al. As seen from Table 3, the total coordination number increases with increase in the concentration of Al i.e., from 70 to 86 atomic percent of Al in Fe-Al alloys and the values changes from 14.10 to 14.84. The obtained results indicate a linear dependence of the K, r and ψ values on the concentration.

The partial nearest-neighbor distances, $r_{(i-j)}$, estimated as the position of the first peak in $g_{ij}(r)$ of the liquid Fe-Al alloys, at 50K above the liquidus temperature are listed in Table 4. It is found that $r_{(Al-Al)}$ has the shortest neighbor distances and $r_{(Fe-Fe)}$ has the longest neighbor distance which follows the same trend as the experimental result reported by [8].

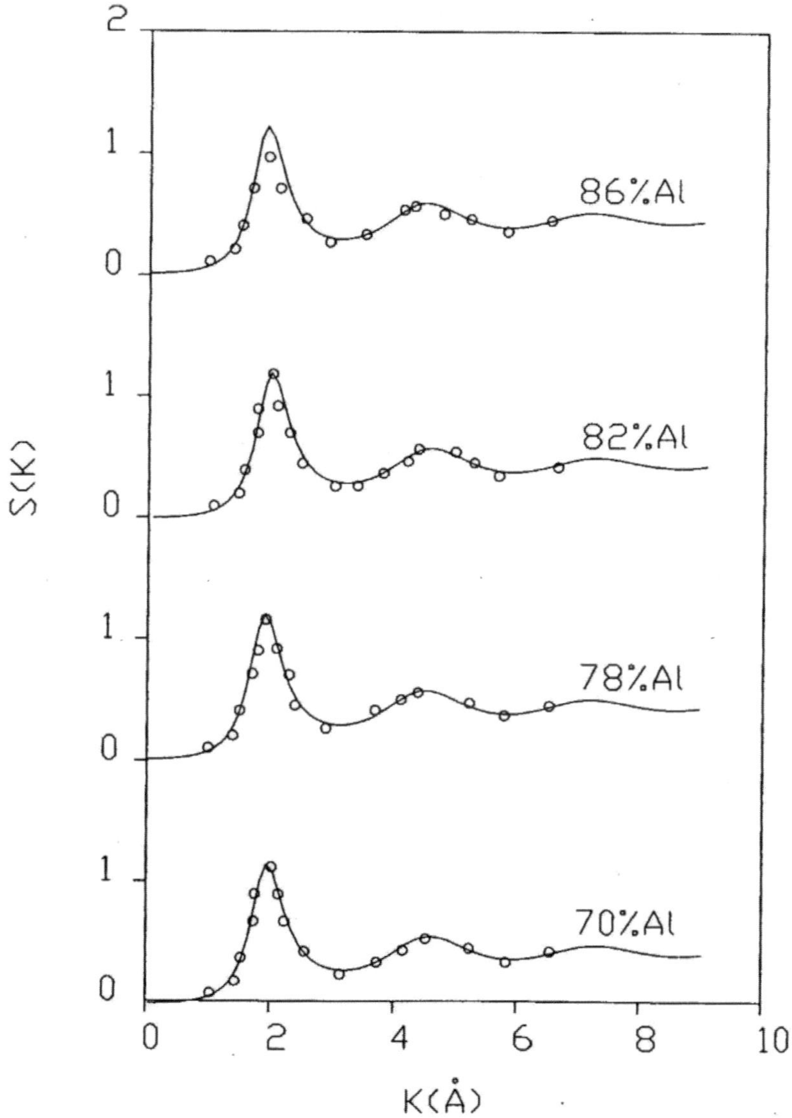

Figure 1. Total structure factor S(k) versus k at different concentrations of Al in Fe-Al alloy; (—) theoretical values, (o o o) experimental values.

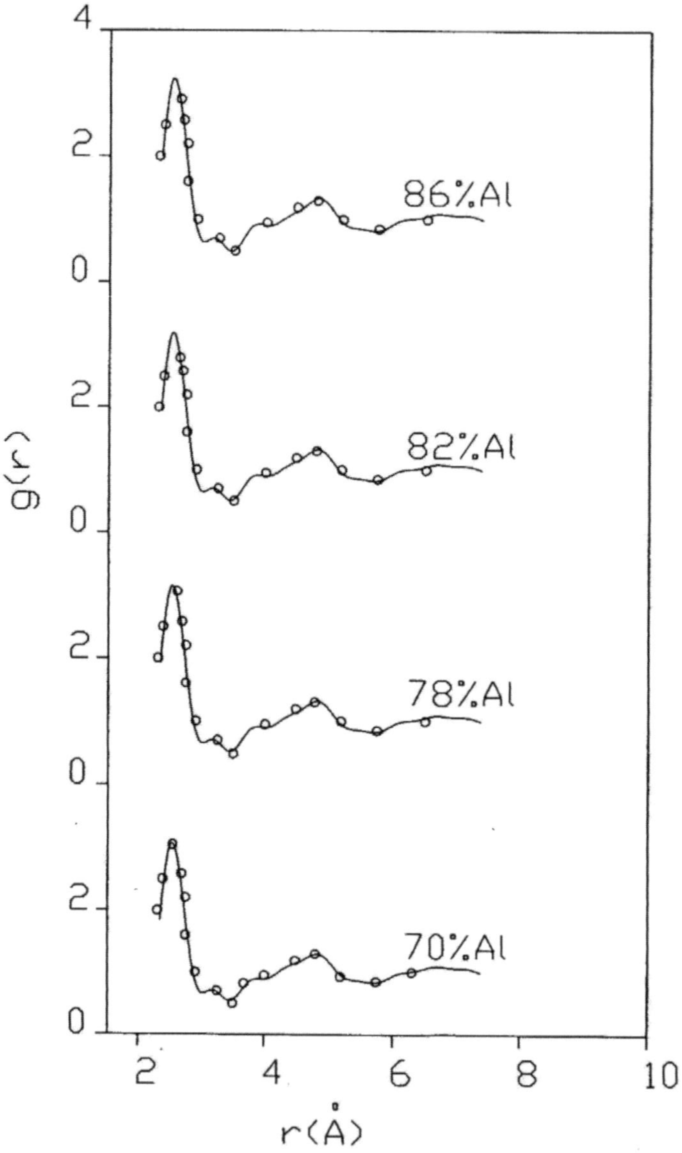

Figure 2. The pair correlation function g(r) versus r at different concentrations of Al in Fe-Al alloy ; (—+— theoretical values, (o o o) experimental values.

4. Conclusions

Structure factors and radial distributions of Fe-Al alloys at different atomic percent of Al were calculated at50K above the liquidus temperature. Computed results were compared with the available experimental values and we found a good agreement between them. The coordination number increases with increase in the concentrations of Al. It was also found that the partial nearest-neighbor distances,$r_{(i-j)}$ shows a reasonable dependence on the concentration of Al atomic percent and was observed to fo0llow the same trend as the experimental results. Therefore, perturbation of Square well attractive tail over hard sphere reference system serves well for the study of static structure factor and derived properties of liquid binary alloys.

Acknowledgment
R K Mishra acknowledges the financial support from the Department of Science and Technology, Govt. of India, New Delhi under Fast Track Project (SR/FT/CS-021/2008) and C. Lalnuntluanga to NFST-UGC for financial assistance.

References
[1] Venkatesh R, Mishra R K and GopalaRao R V 2003 *Phys. Stat. solidi b* **240(3)** 549
[2] Milekhine V, Onsoien M, Solberg J and Skaland T 2002 *Intermetallics* **10** 743
[3] Tsai A P 2003 *Acc. Chem.* **36** 31
[4] Das S, Perepezko J, Wu R and Wilde G 2001 *Mater. Sci. Eng. A* **304-306** 159
[5] Mishra R K and Venkatesh R 2008 *Chem. Phys.* **354** 112
[6] Yokoyama I 2001 *Physica B* **293** 338
[7] Dubinin N E, Filippov V V, Malkhanova O G and Vatolin N A 2009 *Cent. Eur. J. Phys.* **7(3)** 584
[8] Roik O S, Muratov O S, Yakovenko O M, Kazimirov V P, Golovataya N V and Sokolskii V E 2014 *J. Mol. Liq.* **197** 215

Electronic structure and thermoelectric property of Co_2YGe (Y=Mn, Fe) Heusler compounds: a first principle study

Himanshu Joshi[1], D P Rai[2], Sandeep[1] and R K Thapa[1]

[1]Condensed Matter Theory Research Group, Mizoram University, Aizawl, Mizoram, 796004 - India
[2]Pachhunga University College, Aizawl, Mizoram, 796001 – India

Abstract. The electronic and thermoelectric properties of Co_2YGe (Y=Mn, Fe) Heusler compounds have been studied by first principle density functional theory and compared with the known experimental and theoretical results. Results of the density of states (DOS) and band structures shows the half-metallicity of the Heusler alloy Co_2MnGe, whereas the Heusler alloy Co_2FeGe fails to give half-metallicity when treated with GGA. The ZT value calculated for these materials is much below the benchmark value 1.

1. Introduction

Heusler thermoelectric materials have been attracting extensive research interest since their discovery in 1903 [1]. Heusler compounds are ternary intermetallic compounds consisting usually of two transition metals (X, Y) and one main group element (Z). They can be divided into two groups: Compounds with chemical formula X_2YZ having four *fcc* sublattices, and compounds with formula XYZ, consisting of three *fcc* sublattices [2].

The full-Heusler compounds commonly appear as metals. However semi-metallic behavior was observed for Co_2MnGe, as evidenced by several theoretical calculations [3, 4]. The presence of pseudogap in Co_2MnGe is predicted to be due to hybridization effects [3]. The pseudogap formation arises from an indirect band overlap at the Fermi energy (E_F) and thus Co_2MnGe is characterized as semi-metal. In this report, we intend mainly to study the thermoelectric properties by calculating the figure of merit *ZT*, which is defined as the maximum efficiency of a thermoelectric material and is given by

$$ZT = \frac{S^2\sigma}{\kappa}T \qquad [5]$$

where, S is the Seebeck coefficient, σ the electrical conductivity, and κ the thermal conductivity. The thermal conductivity κ compromises of two parts, $\kappa = \kappa_e + \kappa_p$ where, κ_e *and* κ_p are respectively the electronic and phononic contribution to thermal conductivity. Results obtained for Co_2FeGe and Co_2MnGe excludes the phononic contribution to thermal conductivity.

2. Crystal structure and computational details

2.1. Crystal Structure

The full-Heusler crystal structure consists of four penetrating fcc sublattices with atoms at X1(1/4,1/4,1/4), X2(3/4,3/4,3/4), Y(1/2,1/2,1/2) and Z(0,0,0) positions which result in the L_{21} structure having space group Fm-3m. The two Co-site *fcc* sublattices combine to form a simple cubic sublattices as shown in figure 1.

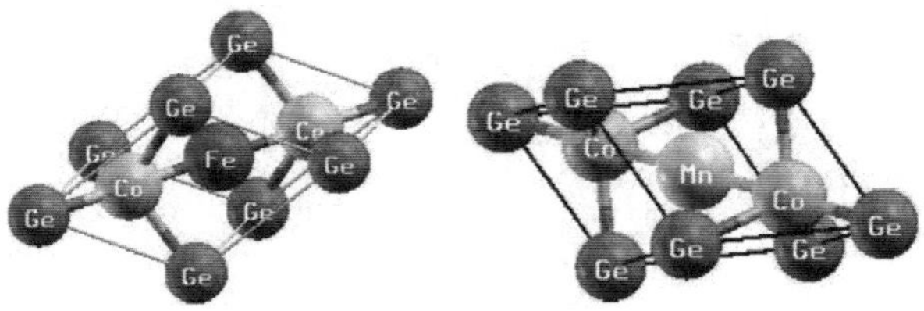

(a) (b)

Figure. 1: (a) Unit cell structure of Co_2FeGe, (b) Unit cell structure of Co_2MnGe

2.2. Computational Details

The total energy is calculated by using Generalized Gradient Approximation (GGA) [6] and Linearized Augmented Plane Wave method (LAPW) [7], the electronic and thermoelectric properties were calculated using Full Potential Linearized Augmented Plane Wave method (FP-LAPW) [8] as implemented in the code WIEN2k [9]. The convergence of basis set was controlled by a cutoff parameter R_{MT} x K_{max} = 7, where R_{MT} is the smallest of the MT sphere radii and K_{max} is the largest reciprocal lattice vector used in the plane wave expansion and made the expansion up to l_{max} = 6 in the muffin tins, where l_{max} is the maximum value of the angular momentum. The magnitude of the largest vector in charge density Fourier expansion (G_{max}) was 12 a.u^{-1}. The cutoff energy which defines the separation of valance and core states was chosen as -6.0 Ry. For k-point sampling, a 31x31x31 k-point mesh in the first Brillouin zone was used resulting in 816 k-points in the irreducible part of the Brillouin zone. The convergence criterion for self-consistence calculations was set up to charge convergence equal to 10^{-3} e^- and the energy convergence criterion was set to 10^{-5} Ry. The set of valance orbitals in the calculations were selected as *3s, 3p, 4s, 4p, 3d* for Co atoms, *3s, 3p, 4s, 4p, 3d* for Mn atoms, *3d, 4s, 4p* for Ge atoms and *3s, 3p, 4s, 4p, 3d* for Fe atoms. All lower states were treated as core states.

A program code called BoltzTraP based on Boltzmann semi-classical transport equation was used to calculate the electron transport properties as implemented in a code BoltzTraP [10]. This code depends on the data provided by WIEN2k code.

3. Results

The volume optimization was performed using the lattice constants taken from the experimental results. The calculated total energies with GGA as a function of the volume were used for determination of theoretical lattice constant and Bulk modulus. The Bulk modulus was calculated using the Murnaghan's equation of state [11]. The calculated value of lattice constant and bulk modulus are presented in Table 1 and the volume optimization curve is shown in figure 2. For Co_2MnGe, the calculated Bulk modulus is 216.6253 GPa, its pressure derivative is 9.3865 the optimized lattice constant is 5.6748 Å and the change in the lattice constant with that of experimental one is 0.0729. Similarly, for Co_2FeGe the calculated Bulk modulus is 188.8626 GPa, its pressure derivative is 5.1122, the optimized lattice constant is 5.7469 and the change in the lattice constant with that of experimental one are 0.0039. The change in the lattice parameters is given by Δ (a_o).

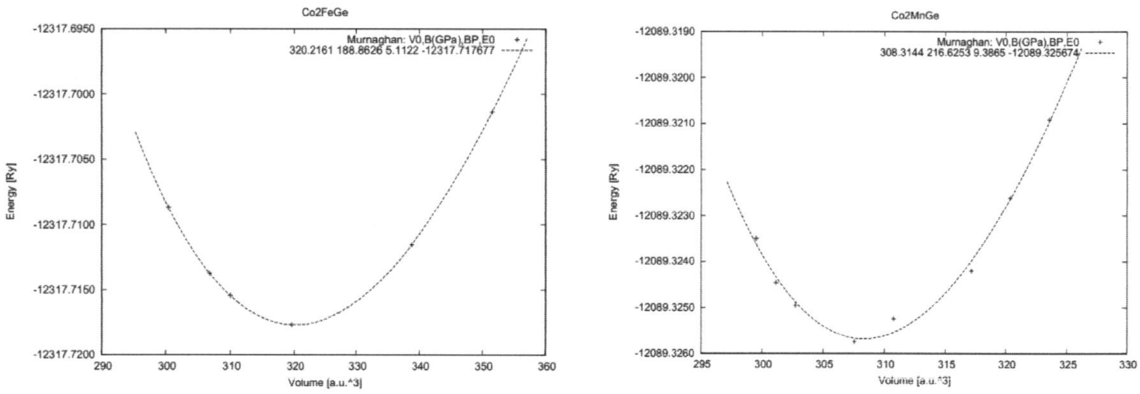

(a) (b)

Figure 2: (a) Volume optimization of Co_2FeGe, (b) Volume optimization of Co_2MnGe

Table 1. Experimental and theoretical lattice constants

Compound	Experimental lattice constant	Calculated lattice constant	Δ (a_o)	Bulk Modulus B (GPa)	Equilibrium Energy (Ry)
Co_2MnGe	5.7477 [12]	5.6748	0.0729	216.62	-12317.717
Co_2FeGe	5.7430 [12]	5.7469	0.0039	188.86	-12089.325

3.1. Density of States (DOS)

The total and partial DOS plots of Co_2MnGe are shown in figures 3 and 4 respectively.

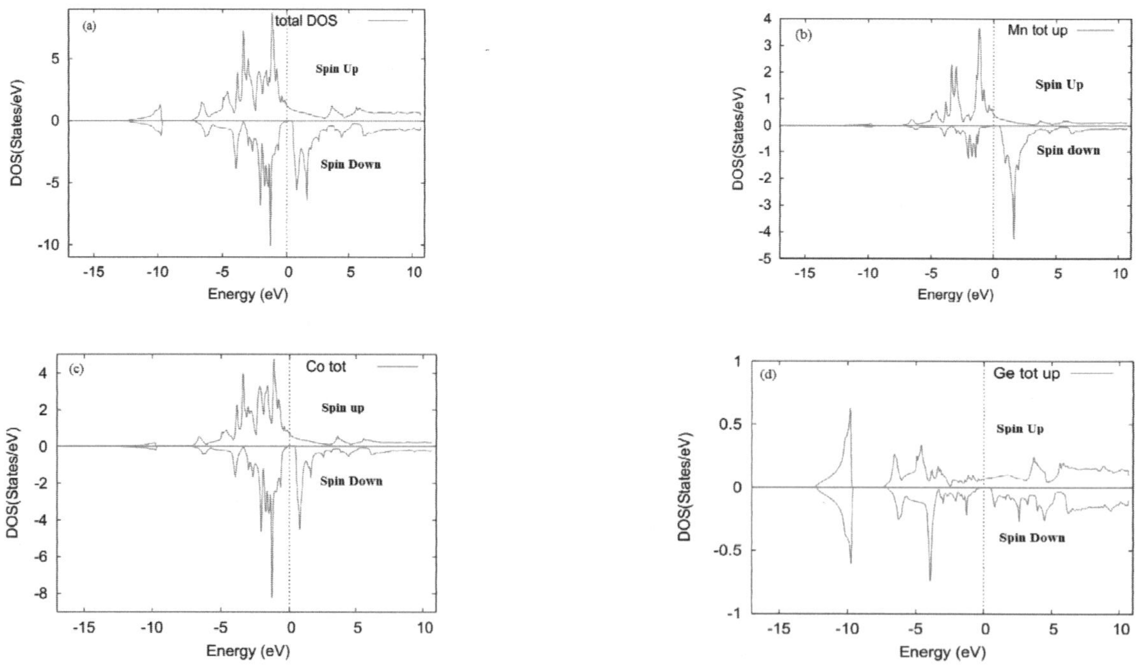

Figure 3. DOS plots of Co_2MnGe. (a) Total DOS plot of Co_2MnGe (b) partial DOS plot of Mn (c) partial DOS plot of Co and (d) partial DOS plot of Ge.

It shows that the maximum contribution in total DOS is due to the Co atom in both spin up and spins down the channel (~ 4.9 eV in spin up and ~ 8.2 eV in spin down). The second largest contribution is due to the Mn atom (~ 3.7 eV in spin up and ~ 4.3 eV in spin down), whereas the Ge atom has the least contribution for the total DOS (<1 eV for both spins). The peaks are mostly due to the *d*- state electrons in semi-core and valance region, below E_F for both spin up and down channel. The contribution in Mn DOS is mainly due to Mn-*d* electrons in the valance region with sharp peaks at -2.35 eV and -2.8 eV. In the conduction region, the sharp peaks are observed at 2.1 eV and 0.9 eV respectively, which are mainly due to the d-e_g and d-t_{2g} states of Co and Mn atoms in the spin down channel (figure 4 (a) and (b)). In the spin up the region of Co_2MnGe (figure 3(a)), DOS intersects E_F, giving metallic nature, whereas, in the spin, down channel, DOS intersection to E_F is absent giving semi-metal nature.

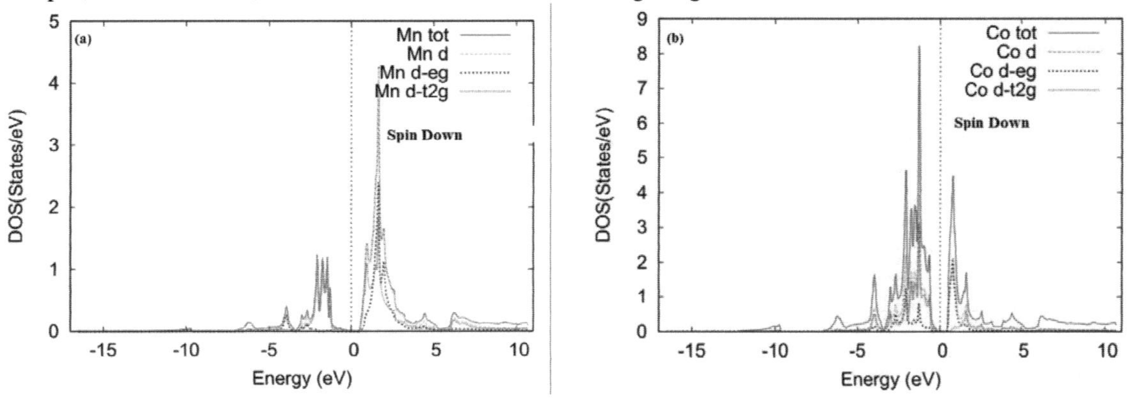

Figure 4. DOS plots of Co_2MnGe. (a) Mn total, *d, d-eg* and *d-t2g* states in spin down and (b) Co total, *d, d-eg* and *d-t2g* states in spin down.

The following figures 5 and 6 gives the total and partial DOS of Co_2FeGe respectively. DOS plot of Co_2FeGe shows that the maximum contribution in total DOS is due to the Co atom (~7.95 eV). The second largest contribution is due to the Fe atom (~5.8 eV), whereas the Ge atom has the least contribution for the total DOS (< 0.6 eV). For Co_2FeGe, the peaks are mostly due to the *d*- state electrons in semi-core and valance region, at E_F (figure 6). The contribution in Fe DOS is mainly due to Fe-*d* electrons in the valance region with sharp peaks at -3.7 eV, -2.2 eV and -2.1 eV and that in Co DOS is mainly due to Co-*d* electrons in the valance region with sharp peaks at -2.2 eV and -2 eV. In the conduction region, the sharp peaks are observed at 1.6 eV and 1.5 eV respectively, which are mainly due to the d-e_g and d-t_{2g} states of Co and Fe atoms.

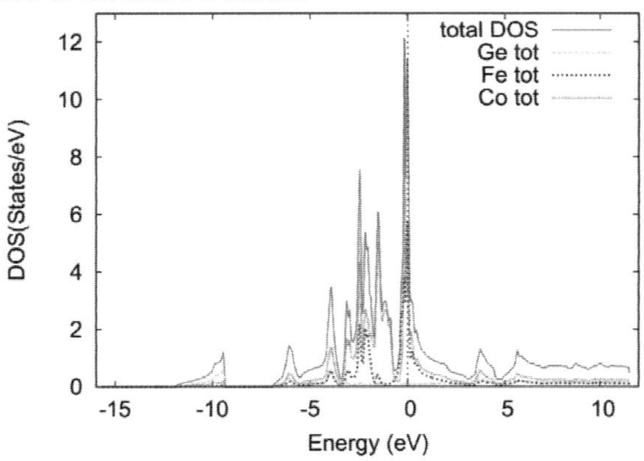

Figure 5. Total DOS of Co_2FeGe

In both Co_2MnGe and Co_2FeGe, the *d*-electrons of Co forms a strong hybridization with Mn-*d* and Fe-*d* electrons.

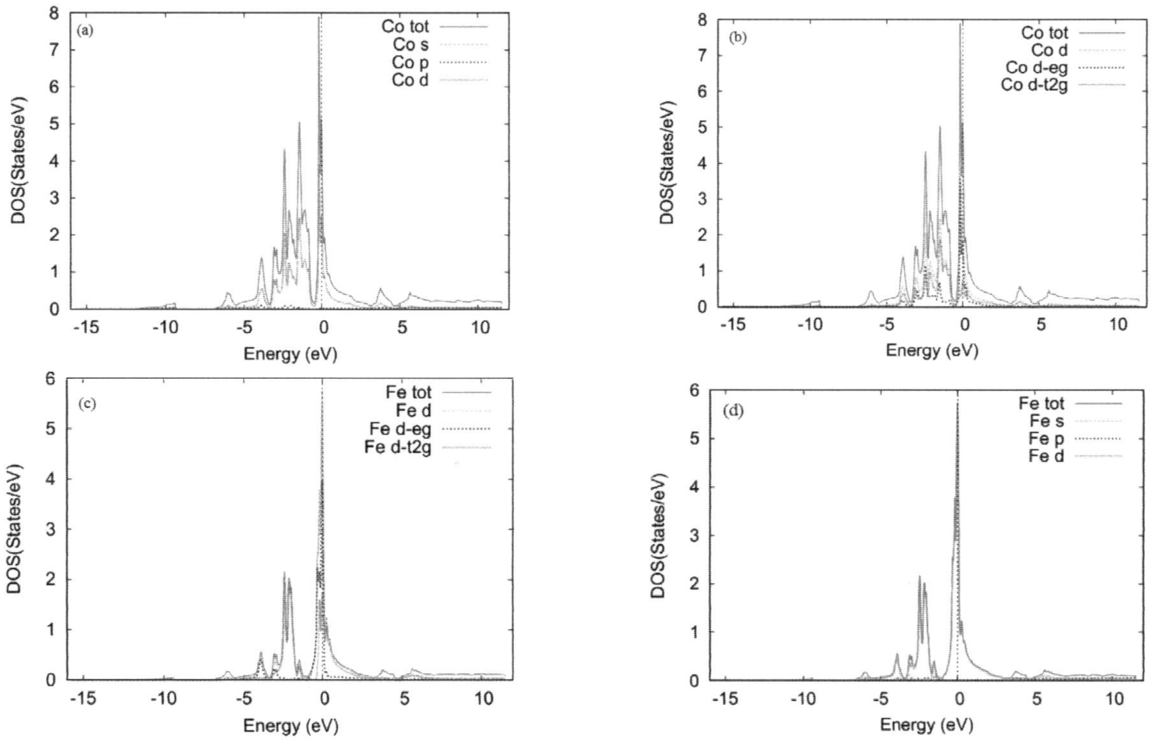

Figure. 6: DOS plots of Co_2MnGe (a), Co total, *s, p,* and *d* states (b) Co total, *d, d-eg* and *d-t2g* states, (c) Fe total, *d, d-eg* and *d-t2g* states and (d) Fe total, *s, p,* and *d* states.

3.2. Band structures

Figures 7(a), (b) and (c) gives the band structures of Co_2MnGe for spin up, spin down channels and for Co_2FeGe respectively. The valance region of the spin up and spin down channel of Co_2MnGe has more number of bands due to the *3d* states of Mn atoms. In the conduction region (spin down), dense bands are observed which is also due to the 3d electrons of Mn atom. In comparison with DOS results, it can be seen that in the valance region, the bands below -3eV consists mainly of the *s* and *p* electrons of the Ge atom for both Co_2MnGe and Co_2FeGe.

In the energy gap between -3.0 eV to 2 eV, the bands are mainly due to the *d* electrons of Co and Mn atoms for Co_2MnGe and *d* electrons of Co and Fe atoms for Co_2FeGe. For Co_2MnGe, from band structure and DOS calculations, it is seen that the band gap originates due to strong hybridization between the *d* states of the higher valent and the lower valent transition metal ions.

Rai *et al.* described that the interaction of Mn with the *Z-p* states splits the Mn-*3d* states into a low-lying triplet of t_{2g} states and a higher-lying doublet of e_g states [6]. The band gap in E_F is formed when the Mn-*3d* states in the majority band is shifted to lower energies, while the Mn *3d* states in the minority band are shifted to higher energies and remain unoccupied. The absence of sharp peaks in the DOS is accounted by a low number of bands in the conduction region of spin-up channel. An indirect band gap of 0.61 eV between symmetry point Γ and *X* was obtained. This value is in close agreement with that obtained by Kandpal *et al.* which was 0.581 eV [13].

Figure 7 (c) gives the energy bands for Co_2FeGe. From the figure we see that Co_2FeGe has no band gap and thus acts as a metal when treated with GGA. However, when treated with GGA+U, Co_2FeGe also acts as a half-metal, giving metallic nature for spin-up channel and semi-metal characteristics for a spin down channel [14].

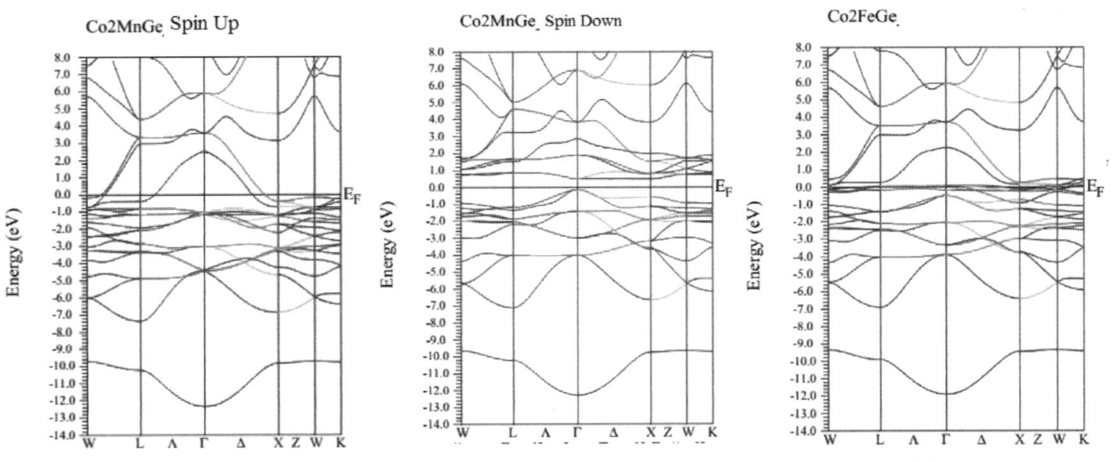

Figure 7. Energy bands of (a) Co_2MnGe Spin up (b) Co_2MnGe Spin down and (c) Co_2FeGe

However, when treated with GGA+U, Co_2FeGe also acts as a half-metal, giving metallic nature for spin-up channel and semi-metal characteristics for a spin down channel [14].

3.3. Thermoelectric properties

The value of S calculated with the proposed method at the room temperature for Co_2MnGe is 16.7×10^{-6} V/K (Figure 8(a)). We find from our study that this value is in agreement with that obtained by Ouardi which was -16×10^{-6} *V/K* [12]. The negative sign in S indicates n – type carriers. However, the value of S calculated for Co_2FeGe is very small. This is because of the absence of band gap in the band structure of Co_2FeGe.

The variation of the relaxation time τ dependent electrical conductivity with temperature is shown in Fig. 8 (b). The electrical conductivity and the power factor have an impact of τ, unless an accurate numerical value of τ is known, the accurate resistivity and power factor (σs^2) cannot be calculated. The electrical conductivity has high values for lower temperatures and for temperatures above 500K, it almost remains constant. This is because of the low density of states at E_F. The value of σ / τ obtained for room temperature was 2.92×10^{22} $(\Omega\ ms)^{-1}$ and 3.37×10^{22} $(\Omega\ ms)^{-1}$ respectively for Co_2MnGe and Co_2FeGe.

The variation of temperature dependent electronic thermal conductivity (κ_e) is shown in Fig. 8(c). The thermal conductivity increases linearly with the rise in temperature above 40 K for both Co_2MnGe and Co_2FeGe. This is in agreement with the results obtained by Ouardi [16]. For lower temperatures, thermal conductivity decreases with the increase in temperature, whereas according to Ouardi, for lower temperatures it increased according to T^3 law. Fig. 8 (d) gives the variation of the figure of merit (maximum performance) with temperature. The *ZT* value calculated at room temperatures are 0.008 and 0.00022 for Co_2MnGe and Co_2FeGe respectively. The value will further reduce if the phononic contribution to thermal conductivity is included in the calculation.

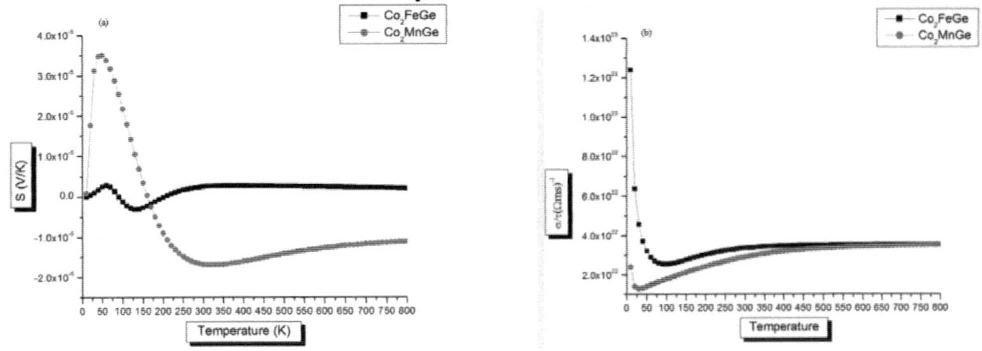

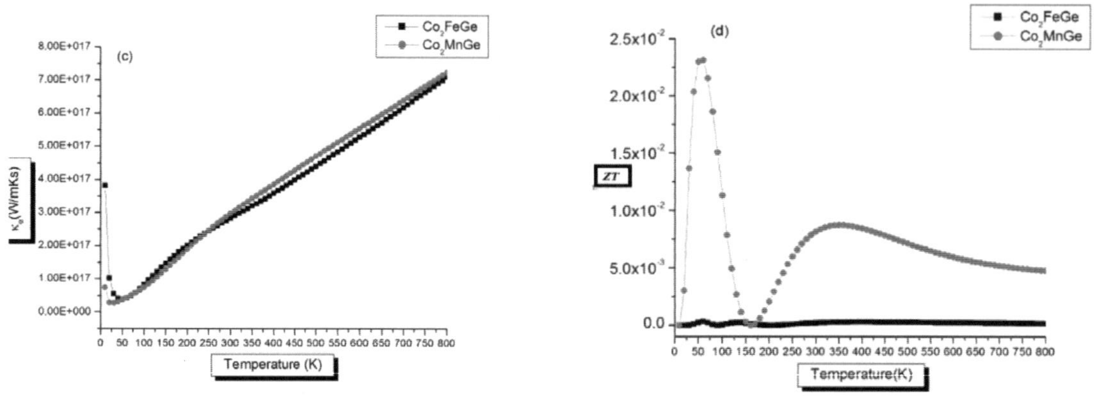

Figure 8. (a)Temperature dependence of Seebeck coefficient (S), (b) electrical conductivity over relaxation time (σ/τ), (c) electron thermal conductivity (k_e) and (d) figure of merit (ZT).

4. Conclusions

We have performed a detailed investigation of the electronic and thermoelectric properties of Co_2MnGe and Co_2FeGe Heusler compounds. The theoretically calculated transport properties are in agreement with the experimental results except for low-temperature regions around 50K. Although the obtained ZT value is much less than the benchmark value for thermoelectric applications, it may be enhanced by doping, nanostructuring etc which attributes to the reduction of thermal conductivity and enhancement of Seebeck coefficient. Both these compounds exist in a half-metallic state indicating that half-metal compounds are poor candidates for thermoelectric applications under normal conditions.

Acknowledgment
HJ would like to acknowledge SERB (India) for project fellowship. RKT acknowledges research grant from SERB (India) via No.EMR/2015/001407

References

[1] Heusler Fr 1903 *Verh. d. DPG* **5** 219.
[2] Graf T, Felser C and Parkin S S P 2011 *Progress in Solid State Chemistry* **39** 1-50
[3] Rai D P and Thapa R K 2012 *Phase Transition* **iFirst** 1-11
[4] Singh D J and Mazin I I 1998 *Phys. Rev. B.* **57** 14352
[5] Kanatzides M G, Chung D Y and Sootsman J R 2009 *Angew. Chem. Inst. Ed.* **48** 8616-8639
[6] Perdew J, Burke K P and Ernzerhof M 1996 *Phys. Rev. Let.* **77** 3865
[7] Hyberstsen M, Schluter M S and Christensen N E 1989 *Phys. Rev. B* **39** 9028
[8] Wimner E, Krasauer H, Wieinert M and Freeman A J 1981 *Phys. Rev. B* **24** 864
[9] Blaha P, Schwarz K, Madsen G K H, Kvasnicka D, Luitz J 2014 *An augmented plane wave + local orbital program for calculating crystal properties* Technische Universitát, Wien
[10] Madsen G K H and Singh D J 2006 *Comput. Phys. Commun.* **175** 67
[11] Murnaghan F D 1944 *Proc. Natl. Acad. Sci. USA* **30** 244
[12] Ouardi S 2012 Electronic structure and Physical properties of Heusler compounds for Thermoelectric and Spintronic Applications. (*Unpublished doctoral thesis*). Mainz University, Germany
[13] Kandpal H C, Fecher G H and Felser C 2007 *Jon. Phys. D* **40** 1507
[14] Rai D P, Shanker A, Sandeep, Ghimire M P and Thapa R K 2012 *Physics B*: **407** 3689-3693

First-Principles Study of Adsorption of Halogen Molecules on Graphene-MoS_2 Bilayer Hetero-system

S. Lamichhane[1], P. Lage[1], G. B. Khatri[1], N. Pantha[1], N. P. Adhikari[1*] and B. Sanyal[2]

[1]*Central Department of Physics, Tribhuvan University, Kathmandu, Nepal.*
[2]*Division of Materials Theory, Department of Physics and Astronomy, Uppsala University, Sweden.*

*email: npadhikari@tucdp.edu.np

Abstract

We have performed density functional theory based first-principles calculations to study the stability, geometrical structures and electronic properties of pure 3×3 supercell of MoS_2, 4×4 supercell of graphene, graphene-MoS_2 bilayer hetero-system, F_2, Cl_2, Br_2 and I_2 molecules on hetero-system within the DFT- D_2 level of approximations. The preferable site and adsorption energy of halogen molecules are studied. The most stable geometries are considered to study their electronic band structure, Density of states and magnetic properties with reference to individual 2D components, graphene and MoS_2 .

1. Introduction

The strongest, thinnest and most stretchable material arranged in honeycomb lattice of thick sheet of sp^2-bonded carbon atoms termed as graphene is at focus of many research works worldwide due to its peculiar properties like observable quantum Hall effect at room temperature and existence of two-dimensional gas of massless Dirac fermions [1-5]. These nobel properties of graphene with its applications in the areas of spintronics, hydrogen storage, sensing [6-9] have attracted the most modern consent of researchers to predict other desirable properties.

On complementary to graphene, MoS_2 belonging to 2D transition metaldichalcogenide, well known as band gap semiconductor is becoming most studied material in the last 5 years [10-12]. MoS_2 is a layered transition metal dichalcogenide semiconductor with an indirect band gap in which Mo and S atoms are stacked together to give S-Mo-S sandwiches coordinated in a triangular prismatic arrangement [13]. In recent time, MoS_2 is attracting the attention of many researchers as it is excessively available in the form of a natural mineral, molybednite [14]. MoS_2 monolayer is widely used to produce transistors [15], integrated logic circuits, signal amplifiers [16], photodetectors and flexible optoelectronic devices. It is a promising material with highly potential applications in solar cells [17], photocatalysts, signal amplifiers and lubrication [18]. However, its properties are limited by band gap. The graphene-MoS_2 bilayer hetero-structure as shown in figure [1] offers the excellent mechanical flexibility, electronic properties, optical transparency, photoconductivity and favorable transport properties [19].

Journal of Physics: Conference Series **765** (2016) 012011 doi:10.1088/1742-6596/765/1/012011

The association of two materials with their distinct electronic properties is believed to drag the desired properties in practical applications. Moreover, the adsorption of foreign atoms on hetrostructure is one of the promising approach to modify and exploit unwanted properties of any constituent. Combination of two materials standing for a brand new family with halogen, commonly known as reactive non metal and highly electronegative elements of periodic table is highly expected to start new direction towards mate-rials science research.Heterostructure may bee important to study the properties beyond the capacities of constituents as well as to eliminate the negative properties and drag desirable one.

Although the graphene-MoS_2 hetero-structure is used to study the adsorption of different metals on nano-scale channels [20], it has not been tested for the ability of heterostructure with reactive non-metals. Previous work for the adsorption of halogens on graphene layers opens very small gap (3-75 meV, highest in fluorine). Since MoS_2 is band gap material, we are interested to test halogens on graphene-MoS_2 bilayer to obtain high band gaps and particularly the role of fluorine to change the electronic structure. From both scientific and technological point of view, investigation of halogen adsorption on semiconducting surface is very important as halogen and its compounds are widely used in the etching technologies. To make comparison for the interaction of halogens and semiconductor surface is of prime importance. In this work, we consider adsorption of diatomic halogen molecules on graphene-MoS_2 bilayer hetero-system.

The remaining part of the paper is organized in the following way. In Sec. II, we discuss the computational details and the systems under our study. Sec. III gives a brief description of results of adsorption of F_2, Cl_2 , Br_2 and I_2 molecules on graphene-MoS_2 bilayer hetero-system. Here, we present the analysis of adsorption sites, binding energies, band structures and density of states. The last conclusion section summarizes the major findings and possible extensions of the ongoing research.

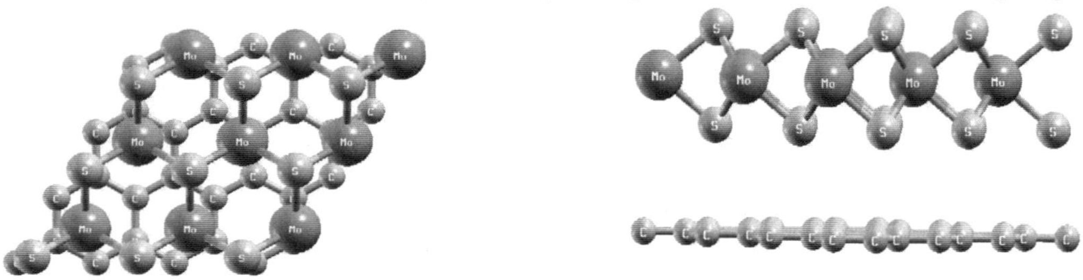

Figure 1: 3×3 supercell of MoS_2 Monolayer and 4×4 super cell of graphene to form a MoS_2 -graphene bilayer hetero-system, (a) Top-view and (b) Side-view respectively

.

2. Computational Details

In the present work, we have incorporated Density Functional Theory (DFT) [21, 22] implemented in the Quantum Espresso (QE) [23] code to study the geometrical and structural properties of 3×3 supercell of MoS_2 , 4×4 supercell of graphene, graphene-MoS_2 bilayer hetero-structure and halogen molecules adsorbed graphene- MoS_2 bilayer hetero-structure. The Perdew-Burke-Ernzerhof (PBE) form of generalized gradient approximation (GGA) [24] with van Der Waals interaction via Grimme's model [25, 26] is used to treat inter electron interaction. The bfgs algorithm with Rappe-Rabe-Kaxiras-Joannopoulos (RRKJ) model of ultrasoft pseudopotential is used to account the interaction between the ion cores and valence electrons. The unit cell is optimized with respect to lattice parameter 'a', kinetic energy cut-off (E_{cut}) for plane wave and the number of k-points along x- and y-axes respectively. Based on these convergence tests, we obtained the lattice constant 'a' for the unit cell of MoS_2 as 3.176 °A which agrees with the previous results [13, 14].

For the 3×3 supercell of MoS_2 the lattice constant is three times that of the unit cell. Further, the plot of the total energy verses the number of k-points reveals that the energy of the unit cell of monolayer MoS_2 is almost constant after n_{kx} equal to 10. Hence a mesh of 10×10×1 k-points, in case of unit cell, could be used for the Brillouin-zone integration. The mesh was reduced to 4×4×1 for 3×3 as per relations of direct and reciprocal lattice geometries, which helps to minimize the computational cost. A plane wave basis set with energy cut-off values of 476 eV (35 Ry) for the wave-function and 4760 eV (350 Ry) for the charge density is used for the expansion of the ground state electronic wave functions.

3. Results and Discussion

3.1 Adsorption of halogen molecules on Graphene-MoS_2 bilayer hetero-structure:

The MoS_2/graphene heterostructures is made using 3 × 3 supercell of MoS_2 and 4 × 4 supercell of graphene by considering the lattice mismatch. There are two possibilities for the construction of the heterostructures accounting the difference in lattice parameters between 18.006 Bohrs of MoS 2 and 18.600 Bohrs of graphene. If lattice constant of MoS_2 is used (i.e MoS_2 is taken reference and graphene is modelled on it), there will be compressive strain of 3.3 % on graphene sheet. If lattice constant of graphene is used i.e graphene as reference and MoS_2 is modelled on it, there will be 3.2 % tensile strain to MoS_2 . In this study, Both cases are taken with the different stacking and configurations to find the minimum energy state for the calculations of the electronic structurs of the heterostructures. Vaccum length is made greater than 20 °A along z- axis to avoid the interactions between the adjacent supercells. We have considered six different occupation sites of high symmetry : the hollow site, at the center of hexagonal plane of MoS_2 , top of molybdenum atom (Mo-Top), top of sulphur atom (S-Top) from MoS_2 side and hollow site at the center of hexagon of graphene (C-Hollow), bridge between two carbon atoms (C-Bridge) and top of carbon atom (C-Top) from graphene side as shown in figure [2]. The adsorbates (halogens) were kept in two different orientations before optimization, viz. Parallel and Perpendicular to the graphene and MoS 2 sheets. Reverse to the order of bond dissociation energy in halogen molecules (Cl_2 >Br_2 >F_2 >I_2), the B.E. of halogens on MoS_2 sheet is found in the order Cl_2 <Br_2 <F_2 <I_2 . The reason is obvious, the halogen molecule with the stronger intra-atomic bonding gain its inert state and becomes less reactive towards MoS_2 monolayer.

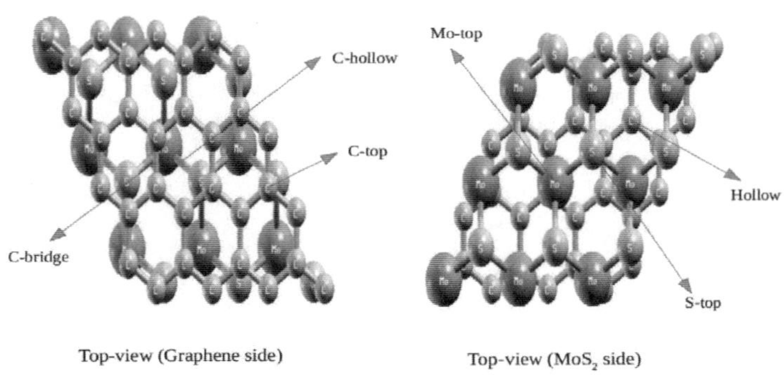

Figure 2: Different possible adsorption sites for the adsorption of halogen molecules on graphene-MoS_2 bilayer hetero-structure.

The binding energy (adsorption energy) of a halogen molecule on graphene-MoS_2 bilayer hetero-system is calculated by using the relation

$$\Delta E = E_{hal} + E_{graphene-MoS2} - E_{hal-graphene-MoS2} \quad (1)$$

where E_{hal} is ground state energy of a fully relaxed halogen molecule, $E_{graphene-MoS2}$ is the ground state energy of graphene-MoS_2 bilayer hetero-structure and $E_{hal-graphene-MoS2}$ is the ground state energy of halogen adsorbed graphene-MoS_2 bilayer hetero-structure. The positive binding energy reflects the stability of the system. The total energy of a halogen molecule adsorbed graphene- MoS_2 bilayer hetero-structure is calculated in 3×3 supercell of MoS_2 and 4×4 supercell of graphene. Out of the six adsorption sites, the site with the largest binding energy is indicated as the most favourable site for adsorption of the halogen molecule. The different parameters obtained from our calcula-tions are shown in tables below.

Table 1: Table for Binding Energy (E), perpendicular distance of centre of adsorbed molecule from sulphur plane of MoS_2 (h_l), reference (h_l)*[27], distortion on MoS_2 sheet (d_{MoS2}) and observed bond length of adsorbed halogen molecule (d) along graphene side of graphene- MoS_2 bilayer heterostructure

Halogen	Orientation	C-hollow					C-bridge					C-top				
		ΔE (eV)	h_l (Å)	h_l^* (Å)	d_{MoS2} (Å)	d (Å)	ΔE (eV)	h_l (Å)	h_l^* (Å)	d_{MoS2} (Å)	d (Å)	ΔE (eV)	h_l (Å)	h_l^* (Å)	d_{MoS2} (Å)	d (Å)
F_2	Parallel	1.882	3.560	3.230	0.119	2.823	0.417	3.700	3.170	0.011	3.010	1.012	3.720	3.230	0.014	1.650
	perpendicular	0.284	3.900	3.700	0.001	2.610	0.234	3.380	3.470	0.008	3.088	0.437	3.390	3.460	0.003	1.640
Cl_2	Parallel	0.163	3.580	3.580	0.035	1.900	0.189	3.590	3.630	0.040	2.000	0.182	3.540	3.640	0.041	2.000
	perpendicular	0.153	4.170	4.340	0.007	2.000	0.188	4.110	4.210	0.005	2.010	0.194	4.180	4.210	0.006	2.020
Br_2	Parallel	0.239	3.610	3.770	0.020	2.310	0.267	3.690	3.740	0.013	2.310	0.258	3.660	3.770	0.030	2.310
	perpendicular	0.225	4.510	4.670	0.002	2.320	0.273	4.320	4.510	0.005	2.330	0.281	4.200	4.450	0.005	2.340
I_2	Parallel	0.365	3.590	3.850	0.019	2.699	0.389	3.520	3.800	0.002	2.695	0.383	3.650	3.830	0.001	2.694
	perpendicular	0.315	4.600	4.730	0.019	2.699	0.352	4.698	4.740	0.017	2.709	0.359	4.750	4.850	0.020	2.709

Table 2: Table for Binding Energy (E), perpendicular distance of centre of adsorbed molecule from sulphur plane of MoS_2 (h_l), distortion on MoS_2 sheet (d_{MoS2}) and observed bond length of adsorbed halogen molecule (d) along MoS_2 side of graphene-MoS_2 bilayer heterostructure

Halogen	Orientation	hollow				Mo-top				S-top			
		ΔE (eV)	h_l (Å)	d_{MoS2} (Å)	d (Å)	ΔE (eV)	h_l (Å)	d_{MoS2} (Å)	d (Å)	ΔE (eV)	h_l (Å)	d_{MoS2} (Å)	d (Å)
F_2	Parallel	0.291	3.002	0.001	1.527	0.229	3.990	0.001	1.526	0.199	3.800	0.001	1.527
	perpendicular	0.261	3.600	0.021	1.538	0.284	3.475	0.000	1.532	0.732	3.200	0.079	1.798
Cl_2	Parallel	0.158	3.400	0.060	1.990	0.162	3.370	0.065	1.995	0.156	3.430	0.070	2.000
	perpendicular	0.159	4.040	0.004	2.010	0.150	4.060	0.054	2.010	0.247	3.980	0.008	2.060
Br_2	Parallel	0.238	3.390	0.040	2.310	0.239	3.400	0.040	2.310	0.239	3.250	0.070	2.310
	perpendicular	0.244	4.200	0.003	2.330	0.226	4.220	0.050	2.320	0.404	4.020	0.001	2.330
I_2	Parallel	0.387	3.398	0.003	2.695	0.385	3.394	0.004	2.695	0.379	3.459	0.003	2.694
	perpendicular	0.344	4.445	0.014	2.712	0.325	4.550	0.011	2.706	0.489	4.150	0.024	2.764

Tables 1 & 2 show the binding energy values of different structures at different occupation sites. The perpendicular distance of centre of adsorbed molecule from sulphur plane of MoS_2 (h_1) is found to agree well with the reference values of (h_1) [27]. The bond length of adsorbed molecule on graphene-MoS_2 bilayer hetero-structure is not obtained to be significantly changed during adsorption process because the obtained molecule-surface interaction is much weaker than the intra molecular bonding in halogens. Here, we consider the system with the highest B.E. or the least total energy is most stable one. The fluorine adsorbed graphene- MoS_2 bilayer hetero-structure is most stable at the hollow site (C-Hollow) at the center of hexagon of the graphene sheet in the parallel configuration. But all the remaining halogen molecules adsorbed hetero-structures are obtained to be most stable while they are perpendicular to the sheet and above the sulphur atom of MoS_2 sheet (S-Top). The highest value of B.E. are 1.882 eV, 0.247 eV, 0.404 eV and 0.489 eV respectively for fluorine atom, Cl_2, Br_2 , and I_2 molecules adsorbed hetero-structures. The geometrical structures with the highest B.E. are considered to draw the various properties of the respective systems. In addition, the distance between two F-atoms increases when it is bound more strongly to MoS_2 sheet of graphene- MoS_2 bilayer hetero-structure. This result indicates that the bonding of F or F_2 with the substrate happens on the cost of weakening of F-F interaction. The adsorption geometries of molecular halogens (i.e. perpendicular and parallel orientations) above the graphene side of hetero-structure is shown in figure [3], while the optimized structures of different halogens on hetero-structure at the most stable occupation site are shown in figure [4].

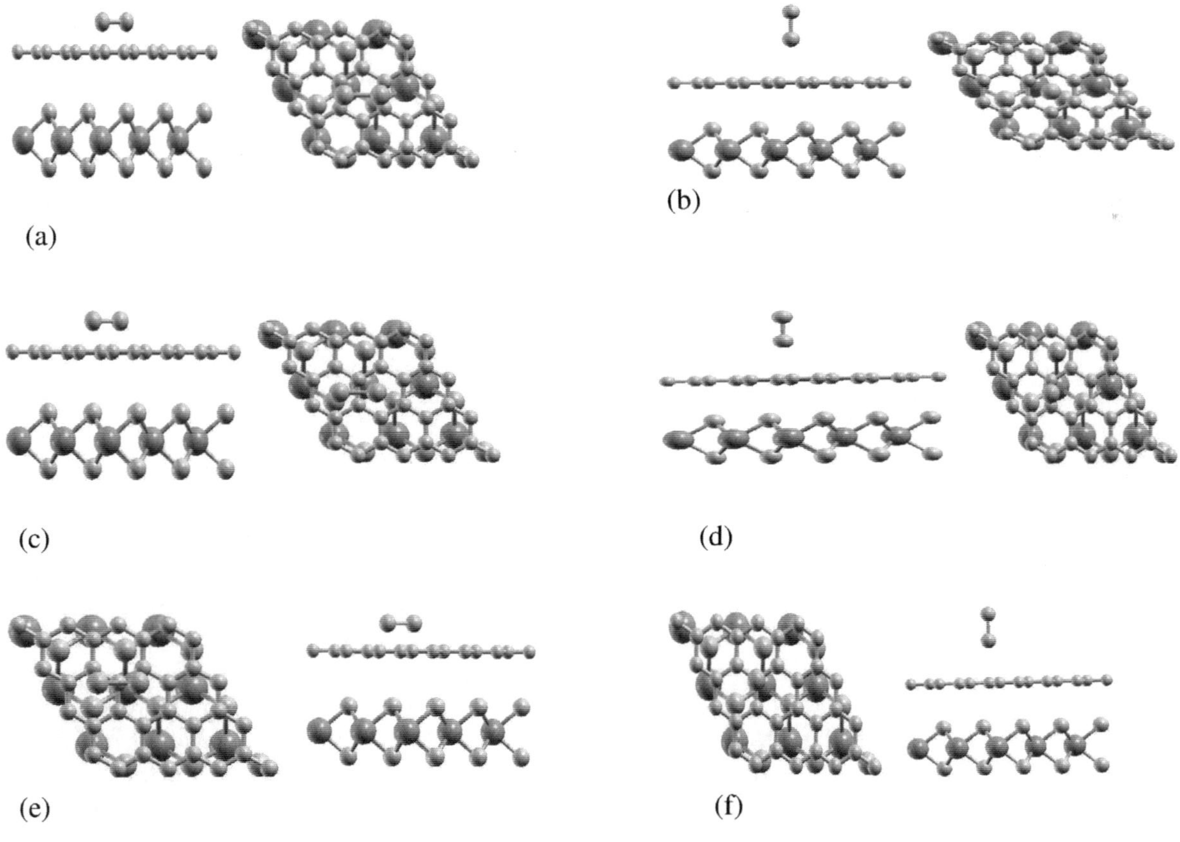

Figure 3: Top and Side view respectively for the adsorption of halogen molecules with different orientations of adsorbates in the direction parallel ((a), (c), (e)) and perpendicular (fig. (b), (d), (f)) to the graphene sheet. Figures (a) and (b) show the adsorption in C-Hollow (H) region, (c) and (d) at C-Bridge and (e) and (f) at C-Top, respectively.
Similarly, the adsorption of halogen molecules with different orientations of adsorbates in the parallel and perpendicular direction to the Mo $_2$ sheet are tested to know the most occupation site.

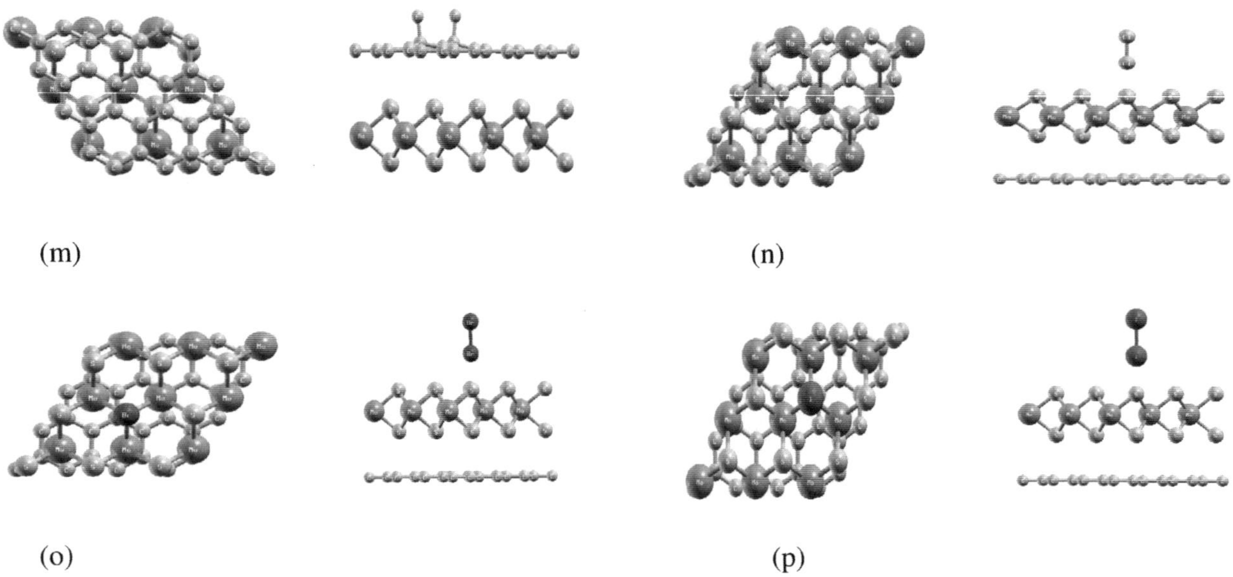

Figure 4: Optimized geometries of a halogen molecule adsorbed graphene-MoS_2 bilayer hetero-structure with top and side-view respectively. The figures, (m) fluorine atoms parallel to graphene plane at Hollow-site (C- hollow), (n) Cl_2 molecule perpendicular to MoS_2 plane at S-Top site (o) Br_2 molecule perpendicular to MoS_2 plane at S-Top site, (p) I_2 molecule perpendicular to MoS_2 plane at S-Top site.

3.2. Electronic structure calculations:

In order to understand the impact of adsorption of halogen molecules, we first need to understand the electronic properties of 3×3 sheet of MoS_2 monolayer as shown in figure [5(a)], 4×4 sheet of graphene as in figure [5(b)] and graphene-MoS_2 bilayer hetero-structure as in figure [5(c)]. Figure [5(a)] shows the band structure of 3×3 sheet of MoS_2 monolayer. Since, the conduction band minimum and valence band maximum lie at the same symmetric point, it is a direct band gap semiconductor with band gap of 1.65 eV. In case of 4×4 sheet of graphene, there is no band gap as conduction band and valence band coincides at Fermi level indicating that the graphene is a zero band gap semiconductor. Figure [5(c)] shows the band structure of graphene-MoS_2 bilayer hetero-structure. Here, the Dirac point shifts about 0.40 eV above the Fermi level which shows that the graphene-MoS_2 bilayer hetero-structure is metallic in nature. Afterwards, we have studied the effect of adsorption of halogen molecules on the hetero-structure along graphene plane and MoS_2 plane. The band structures of different systems are shown in figures [5(d)], [5(e)], [5(f)], and [(5(g).

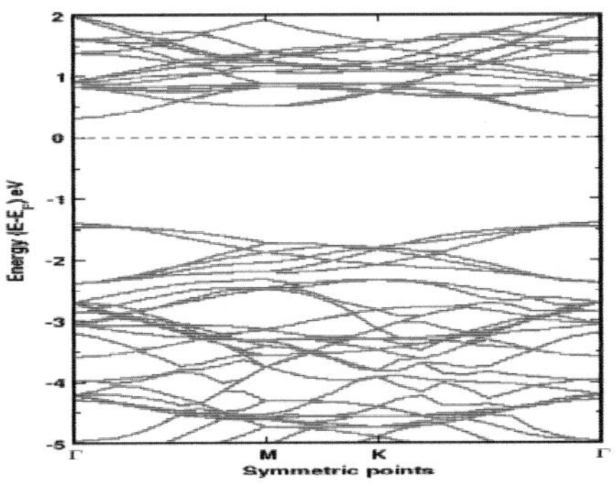
Energy (E-E$_F$) eV
Γ
M
K
Γ
Symmetric points

(a)

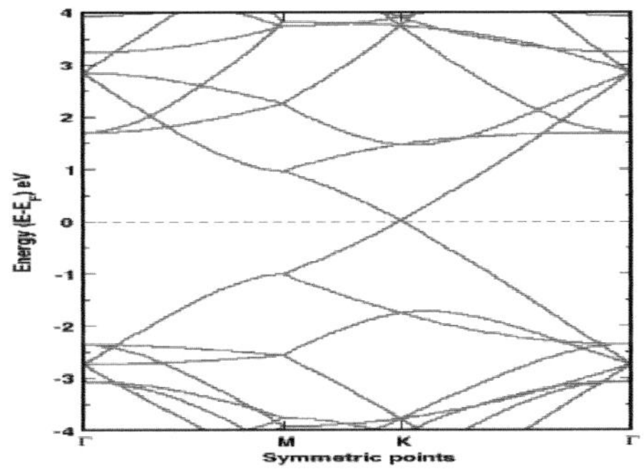
Energy (E-E$_F$) eV
Γ
M
K
Γ
Symmetric points

(b)

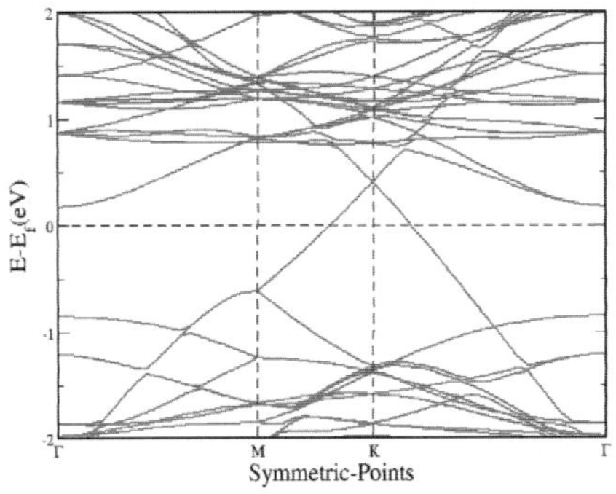
E-E$_f$(eV)
Γ
M
K
Γ
Symmetric-Points

(c)

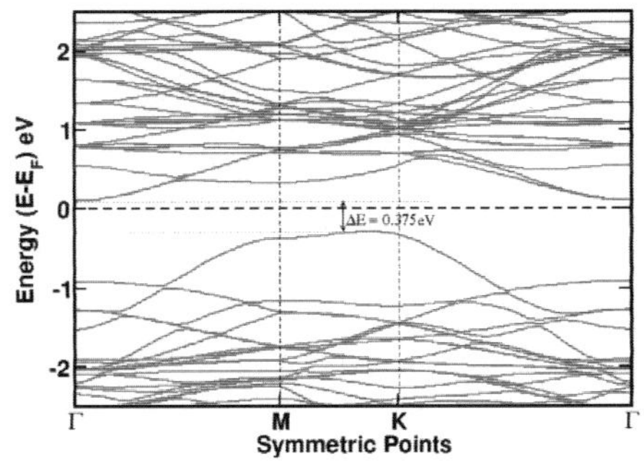
Energy (E-E$_F$) eV
ΔE = 0.375eV
Γ
M
K
Γ
Symmetric Points

(d)

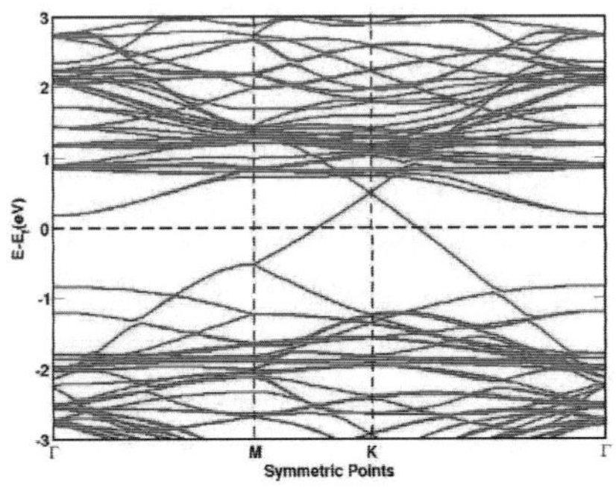
E-E$_f$(eV)
Γ
M
K
Γ
Symmetric Points

(e)

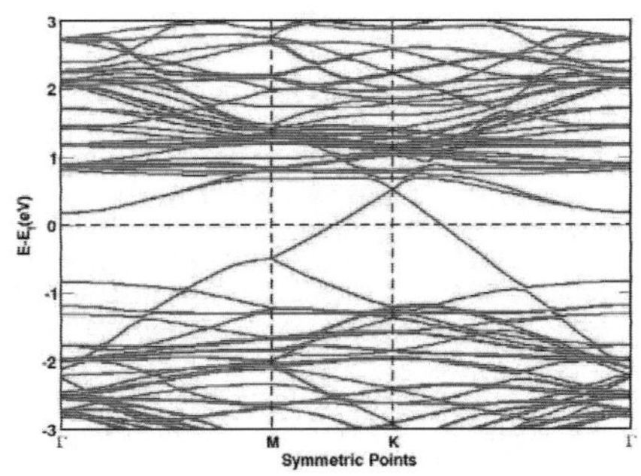
E-E$_f$(eV)
Γ
M
K
Γ
Symmetric Points

(f)

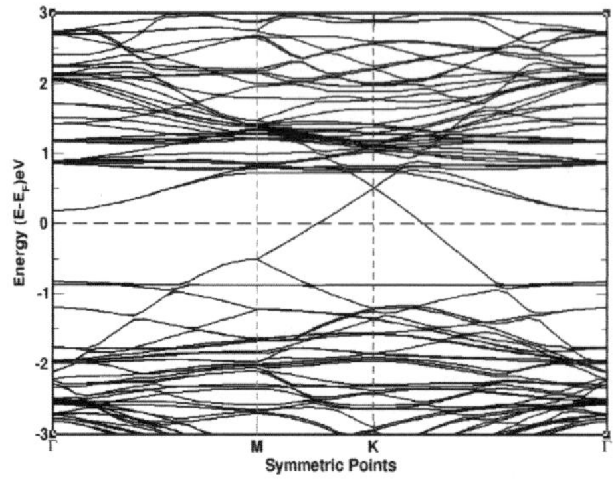

(g)

Figure 5: Band structures of (a) 3×3 sheet of MoS_2 monolayer (b) 4×4 sheet of graphene (c) graphene-MoS_2 bilayer (d) fluorine atoms (e) Cl_2 (f) Br_2 (g) I_2 molecule adsorbed graphene-MoS_2 bilayer hetero-structure.

The modifications in band structures is illustrated in table [3]:

Table 3: Electronic properties of halogen adsorbed Graphene/MoS2 system.

Optimized structure	Band Gap (eV)	Fermi Energy (eV)	Dirac Point(eV)	Dirac Shift (eV)
Graphene-MoS_2 bilayer	-	1.819	0.400	-
F_2-graphene MoS_2	0.375	1.174	-	-
Cl_2-graphene MoS_2	-	1.868	0.480	0.080
Br_2-graphene MoS_2	-	1.870	0.500	0.100
I_2-graphene MoS_2	-	1.996	0.520	0.120

From table [3], it is clearly seen that the band gap of 0.375 eV is found in the case of fluorine atoms adsorbed graphene-MoS_2 bilayer but in other halogen adsorbed bilayer, Dirac point shifted above the Fermi level. The formation of band gap in fluorine adsorbed system is due to the bonding and anti-bonding states of the electrons in the orbitals ans also due to band folding property of electrons in the orbitals. The band structures of other halogen adsorbed systems resembles the band structure of graphene-MoS_2 bilayer. The shift of Dirac point increases from Cl_2 to I_2 adsorbed graphene-MoS_2 bilayer. Also the Fermi energy is seen to be in the increasing order from Cl_2 to I_2 adsorbed graphene-MoS_2 bilayer. Since fluorine atom has the highest electronegative value among halogens, it could easily go for interaction in presence of other reacting species. This could be seen at the B.E. values in table [1]. Our results strongly show the favourable condition to absorb halogens particularly fluorine molecule in excited mode as done to store hydrogen molecule/s on metal decorated graphene [28]. Similarly, the magnetic properties of our system are studied with the help of information displayed by density of states (DOS) as shown in figure [6]

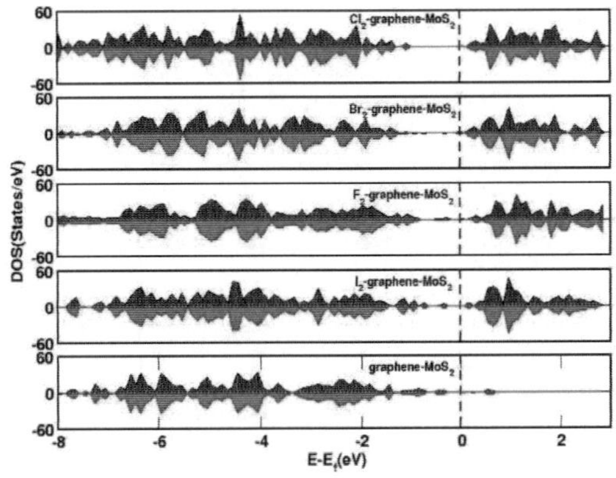

Figure 6: DOS of graphene- MoS_2 bilayer hetero-structure and halogen adsorbed graphene- MoS_2 bilayer hetero-structure. The curve above zero x-axis in each plot represents up-DOS, and that below the axis represents down-DOS.

The plot of DOS shows that both up-DOS and down-DOS are symmetrical about the axis in graphene-MoS_2 bilayer hetero-system as well as halogen molecules adsorbed bilayer hetero-structures. This means, the non magnetic nature of graphene-MoS_2 bilayer hetero-structure is still preserved with adsorption of halogens on it. From figure [6], it is clear that the DOS shift above the Fermi level due to adsorption of halogen molecules on hetero-structure.

4. Conclusions:

We have studied a first-principles study of geometrical and electronic properties of graphene-MoS_2 bilayer hetero-structure and halogen molecules adsorbed hetero-structures. The adsorption sites and orientations are tested to draw the most stable configurations. The fluorine atoms adsorbed system is stable at hollow site of hexagonal plane of graphene (C-Hollow) in parallel configuration with the binding energy 1.882 eV. The Cl_2 , B_2 and I_2 adsorbed hetero-structures all are most stable in the perpendicular configuration above the sulphur atoms of MoS_2 plane (S-Top) with binding energies 0.247 eV, 0.404 eV and 0.489 eV respectively. Band structure unalters in case of Cl_2 , B_2 and I_2 adsorbed graphene-MoS_2 bilayer hetero-structure but Dirac point shifts above the Fermi level. The band gap of 0.375 eV is found in fluorine atoms adsorbed hetero-structure. Our work suggests a way to tune the electronic properties of heterostructure with adsorption of halogens. We then analyzed DOS of respective systems and concluded that monolayer MoS_2 and halogen adsorbed system are non magnetic.

5. Acknowledgements:

The authors acknowledge to Swedish Research Links for partial support.

Journal of Physics: Conference Series **765** (2016) 012011 doi:10.1088/1742-6596/765/1/012011

References

[1] Geim A K and Novoselov K S 2007 *Nature Materials* **6**, 183
[2] Novoselov K S, Geim A K, Morozov S V, Jiang D, Zhang Y, Dubonos S V Grigorieva I V and Firsov A A 2004 *Science* **306**, 666
[3] Zhang Y, Tan Y W, Stormer H L, Kim P, 2005 *Nature* **438**, 201
[4] Zhang Y, Jiang Z, Small J P, Purewal M S, Tan Y W, Fazlollahi M, Chudow J W, Jaszczak J A, Stormer H L and Kim P 2006 *Phys. Rev. Lett.* **96**, 136806
[5] Novoselov K S, Geim A K, Morozov S V, Jiang D, Katsnelson M I, Grigorieva I V, Dubonos S V and Firsov A A 2005 *Nature* **438**, 197
[6] Pak N, Hong S, Kim G and Jhi S H 2007 *J. AM. CHEM SOC.* **129**, 8999
[7] Lin Y, Jenkins M, Valdes-Garcia K A, Small A, Farmer J P, Avouris D P 2009 *Nano Lett.* **9**, 422
[8] Lamichhane S, Pantha N and Adhikari N P 2014 *BIBECHANA* **1**, 113
[9] Zanella, Fagan I, Mota S B and Fazzio R 2008 *A. J. Phys. Chem. C* **112,** 9163
[10] Butler S Z, Hollen S M, Cao L, et al., 2013 ACS *Nano* **7**, 2898
[11] Wilson J A and Yoffe A D 1969 *Adv. Phys.* **18**, 193
[12] Asl M G, Enyashin A N, Kuc A and Heine T 2013 *Phys. Rev. B* **88**, 245440
[13] Kadantsev E S, Hawrylak P 2012 *Sol. St. Comm.* **152,** 909
[14] Lembke D, Bertolazzi S and Kis A, 2015 *Acc. Chem. Res.* **1**, 48
[15] Zhang Y J, Ye J T, Matsuhashi Y and Iwasa Y 2012 *Nano Lett.* **12**, 1136
[16] Radisavljevic B, Whitwick M B and Kis A 2012 *Appl. Phys. Lett.* **101**, 043103
[17] Fornarini L, Stirpe F and Scrosati B 1981 Solar Energy Mater. **5**, 107
[18] Hu K H, Hu X G, Wang J, Xu Y F and Han C L 2012 Tribol. Lett. **47**, 79
[19] Debbichi L, Eriksson O and Lebegue S 2014 *Phys. Rev. B* **89**, 205311
[20] Gelb L D, Gubbins K E, Radhakrishnan R and Sliwinska-Bartkowiak M 1999 *Rep. Prog. Phys.* **62**, 1573
[21] Hohenberg P and Kohn W 1964 *Phys. Rev. B* **136**, 864
[22] Kohn W and Sham L J 1965 *Phys. Rev. B.* **140**, 1133
[23] Giannozzi P, Baroni S, Bonini N, Calandra M, Car R, Cavazzoni C, *et al.* 2009 *J. Phys., Condens. Matter* **21**, 395502
[24] Perdew J P, Burke K, Ernzerhof M 1996 *Phys. Rev. Lett.* **77**, 3865
[25] Grimme S 2004 *J. Comput. Chem.* **25**, 1463
[26] Grimme S 2006 *J. Comput. Chem.* **27**, 1787
[27] Rudenko A N, Keil F J, Katsnelson M I *et. al.* 2010 *Phys. Rev. B.* **82**, 035427
[28] Pantha N, Khaniya A, Adhikari N P 2015 *International Journal of Modern Physics B* **29**, 1550143

Structure, thermodynamic and surface properties of liquid metals investigated by Square well potential

R Lalneihpuii and Raj Kumar Mishra*

Department of Chemistry, School of Physical Sciences,
Mizoram University, Aizawl-796004, India
*Email: rkmishramzu@yahoo.com

Abstract: The auto correlation function, $S(k)$ (where $k = \frac{4\pi}{\lambda} sin\theta$) and its Fourier component g(r) are important quantities characterizing the structure of a liquid. Experimentally these quantities have been determined using neutron or X-ray scattering intensities. Wertheim's solution of the fundamental statistical mechanical equation given by Percus and Yevick for hard spheres mixture is invoked with square well attractive part as a perturbation tail to evaluate the direct correlation function C(k) in momentum space through which structure factor, S(k) of liquid metals were derived. Radial distribution function, g(r) is obtained by Fourier analysis of computed S(k) from which the coordination numbers of ten liquid metals were computed. Surface tension of atomic liquids in terms of diffusion coefficient has been determined under square well interaction. The computed values of structural dependent coordination number, diffusion coefficient and surface tension of these liquid metals are compared with the available experimental results and a good agreement is found between the computed and experimental results

Keywords: Square well potential, structure factor' Diffusion coefficient, coordination Number, surface tension

1. Introduction

The subject of statistical mechanics has been successfully applied in various fields including the prediction of the microscopic function, $S(k)$ of liquid metals [1-3]. A detailed knowledge of the S(k) or g(r) is essential for a quantitative understanding of the structure of liquids and is also sufficient to determine numerous other equilibrium and non equilibrium properties [4].Wertheim [5] and Thiele [6] (WT) solved Percus -Yevick's (PY) equation for hard sphere fluids to obtain the hard sphere direct correlation function $C_{hs}(r)$. Liquid metals static structure factors behave like hard sphere fluids and calculations for thermo-physical and thermodynamic properties with such reference system have been found to be reasonable in many cases [7]. However, the hard sphere reference system lacks realistic properties because thermodynamics and the relation between thermodynamics with $S(k)$ or $g(r)$ are different for hard sphere and real fluids [8]. Hence, it is important to include attraction between the particles in deriving structure factor of liquids. Further, it is pointed out that the success of any theoretical model depends on its experimental confirmation [9, 10]. Thus in this work the square-well (SW) attractive tail has been perturbed over hard sphere reference system (WT) to evaluate the direct correlation function $C(k)$ in momentum space in order to compute $S(k)$ of real liquids. The analytical solution of this model in the mean spherical model was introduced by Rao *et al.* [11- 13]. The model is applied to set of liquid metals to obtain microscopic structural characteristics with their application in the determination of various properties of the considered systems.

The SW potential includes both repulsive and attractive parts and easy to solve numerically and hence it is most suitable for different theoretical techniques, such as integral equations or perturbation theories. The SW potential has been successfully applied for metallic liquids [3, 11-14], colloidal particles [15, 16], hetro-chain molecules [17, 18] and complex systems [19, 20].Here we consider that the multi-particle interactions are due to the sum of pair wise interactions.

The study of surface tension of melts is of fundamental importance and the better understanding of these properties is helpful in material processing technology. Several authors have reported that the surface properties of liquids are very much depending on their bulk micro structural characteristics and their transport properties.

In this work, self–diffusion coefficient, D, of liquids were determined using microscopic structure $S(k)$, $g(r)$ and pair wise interaction SW. Obtained D of considered liquids were employed in the determination of surface tension, γ_{ST}, recently published by many authors with different approaches [21, 22]. Lu *et al.* [23] reported that the γ_{ST} is not well known experimentally even for many simple metals.

2. Theoretical formalism

2.1. *Evaluation of static structure factor and coordination number*

The SW is an extension of hard sphere potential as it retains hard sphere repulsive properties but allows the particles to attract one another and the interaction energy $U(r)$ between two square well particles separated by a distance r is given by

$$\beta U(r) = \begin{cases} \infty & ; \quad r<\sigma \\ -\varepsilon & ; \quad \sigma < r < \lambda\sigma \\ 0 & ; \quad r > \lambda\sigma \end{cases} \tag{1}$$

where, σ is the hard core diameter, $\sigma(\lambda\text{-}1)$ and $\varepsilon(<0)$ are the breadth and depth of the potential well,$\beta = 1/k_B T$.

An important model system is the MSMA and was first proposed by Lebowitz and Percus [22] which expressed $g(r)$ and the direct correlation function $C(r)$ as

$$\left.\begin{aligned} g(r) &= 0 & ; \quad & r<\sigma \\ C(r) &= -U(r)/k_B T & ; \quad & r>\sigma \end{aligned}\right\} \tag{2}$$

We write the Ornstein-Zernike direct correlation function (DCF) of square well fluid in momentum space under MSMA as

$$C(k) = C_{hs}(k) + C_{sw}(k). \tag{3}$$

With

$$\begin{aligned} \rho C_{hS}(k) = &- [24\eta/(x)^6][\alpha(x)^3\{\sin(x\sigma x - x\cos(x)\} + \beta(x)^2 \\ &\{2x\sin(x) - (x^2-2)\cos x - 2\} + \gamma\{(4x^3 - 24x) \\ &\sin(x) - (x^4 - 12x^2 + 24)\cos x - 24\}]. \end{aligned} \tag{4}$$

$$\rho C_{SW}(k) = [24\eta\varepsilon/k_B T]\ (x)^3\ [\sin(\lambda x) - \lambda x\cos(\lambda x) + x\cos(x) - \sin(x)]. \tag{5}$$

where, $x = k\sigma$, ρ is the number density, ε and λ represent depth and breadth respectively, of the square well and the other terms that enter in Eq. (4) are given by following expressions [24].

$$\alpha = \frac{(1+2\eta)^2}{(1-\eta)^4} \tag{6}$$

$$\beta = -\frac{6\eta\,(1+\eta/2)^2}{(1-\eta)^4} \tag{7}$$

$$\gamma = \frac{\eta\alpha}{2} \tag{8}$$

Where η is called packing fraction i.e. volume occupied by the atoms divided by total volume and is given by

$$\eta = \frac{\pi\rho\sigma^3}{6}. \tag{9}$$

The $S(k)$ of one component liquid can be given in terms of $C(k)$ as

$$S(k) = [1 - \rho C(k)]^{-1}. \tag{10}$$

The Fourier inversion of $S(k)$ gives the radial distribution function, $g(r)$.

$$g(r) = 1 + \frac{1}{2\pi^2\rho}\int_0^\infty k^2[S(k) - 1]\frac{\sin(kr_{nm})}{kr_{nm}}dk. \tag{11}$$

Further we derived self-diffusion coefficients of all the concerned liquid metals using well known Einstein's formula $D = \frac{k_B T}{\xi_H + \xi_S + \xi_{HS}}$. Here ξ_H, ξ_S and ξ_{HS} are the friction coefficients due to hard, soft and hard-soft part of the potential function respectively and are obtained in line with Ref. [4] under Helfand's linear trajectory principle using a SW tail as a soft perturbation over hard sphere reference system and can be given as follows

$$\xi_H = \frac{8}{3}\rho\sigma^2(\pi m k_B T)^{1/2} g^{HS}(\sigma) \tag{12}$$

$$\xi_S = -\frac{1}{3}\frac{\rho}{4\pi^2}\left(\frac{\pi m}{k_B T}\right)^{1/2}\int_0^\infty \frac{1}{\rho}k^3 U^{SW}(k)[S(k) - 1]dk. \tag{13}$$

$$\xi_{HS} = -\frac{1}{3}\rho g^{HS}(\sigma)\left(\frac{m}{\pi k_B T}\right)^{1/2} \times \int_0^\infty [k\sigma\cos(k\sigma) - \sin(k\sigma)]\, U^{SW}(k)dk. \tag{14}$$

Here $U^{SW}(k)$ is the Fourier transform of SW potential

$$U^{SW}(k) = \frac{4\pi\varepsilon}{k^3}[Ak\sigma\cos(Ak\sigma) - \sin(Ak\sigma) - k\sigma\cos(k\sigma) + \sin(k\sigma)] \tag{15}$$

2.2. *Relation of self-diffusion coefficient in bulk liquid metals to surface properties.*

The detail studies of surface properties of condensed matter help in understanding their metallurgical processing. The surface tension of elemental liquids can be given by statistical mechanical approach under zeroth order approximation as [25]

$$\gamma_{ST} = \frac{\pi\rho^2}{8}\int_0^\infty \frac{d\,U(r)}{dr} g(r) r^4 dr. \tag{16}$$

Born and Green [26] derived the coefficient of viscosity of liquid metals using statistical mechanical approach as

$$\eta_V = \frac{2\pi\rho^2}{15}\left(\frac{m}{K_B T}\right)^{1/2}\int_0^\infty \frac{dU(r)}{dr} g(r)\, r^4 dr. \tag{17}$$

Here m is the atomic mass of the liquid metals. Eqs. (16) and (17) were derived on a strong scientific basis for hard sphere model but it is not easy to get the numerical solution of the integral equations. A striking result can be obtained using Eqs. (16) and (17) with the well-known Stokes-

Einstein relation, $\eta_V = k_B T/(2\pi a D)$, here a, the nearest neighbor distance, can be taken as first peak position of $g(r)$ for real liquids as [27]

$$\gamma_{ST} = \frac{(k_B T)^{3/2}}{\mathrm{m}^{1/2}} \frac{15}{32\,\pi\, a\,\mathrm{D}}. \tag{18}$$

Since, D can be evaluated from well-known Einstein's relation using the SW long range interaction and hence the surface tension of liquid metals is obtained through Eq. (18).

3. Results and discussion

The WT solution of PY hard sphere fluid within the MSMA with SW perturbation can be solved numerically with the input data: temperature, T, atomic density, ρ, and the SW parameters. The input parameters of liquid metals are listed in Table 1.

TABLE 1. Input parameters of liquid metals with σ as the diameter, ε/k_B as the depth, λ as the breath of the square well potential and ρ as the number density.

Metals	Temperature (K)	σ (nm)	ε/k_B(K)	λ	ρ (10^{25} m^{-3})
Sodium	378	0.330	111.60	1.65	2430
Potassium	343	0.411	96.14	1.65	1276
Cesium	303	0.481	109.5	1.70	0813
Magnesium	953	0.275	127.82	1.43	3900
Aluminum	943	0.245	160.00	1.30	6459
Indium	433	0.283	173.76	1.70	3686
Lead	613	0.297	70.00	1.4	3099
Silver	1273	0.260	500.00	1.75	5159
Copper	1423	0.225	300.00	1.68	7408
Gold	1373	0.260	600.00	1.73	5271

The peak positions and peak heights of the computed $S(k)$ and $g(r)$ for all the considered liquid metals are presented in Table 2 with their experimental data, which give the detailed information regarding the principal peak and structural characteristics of the metals. From Table 2, it is seen that the agreement between our computed $S(k)$ and the experimental results [28] around the first peak appears to be very good.

TABLE 2. Theoretical and experimental values of first positions (k) and peak heights $S(k)$; first peak positions (r) and peak heights $g(r)$ of liquid metals.

Metals	First Peak of $S(k)$				First Peak position $g(r)$			
	Theoretical		Experimental		Theoretical		Experimental	
	k (nm^{-1})	$S(k)$	k (nm^{-1})	$S(k)$	r (nm)	$g(r)$	r (nm)	$g(r)$
Na	20.0	2.7	20.0	2.7	0.36	3.24	0.37	2.42
K	16.0	2.6	16.0	2.6	0.45	2.80	0.46	2.35
Cs	14.0	2.5	14.0	2.7	0.50	3.61	0.51	2.58
Mg	25.0	2.5	24.0	2.5	0.29	2.84	0.31	2.46
Al	27.0	2.4	27.0	2.4	0.27	2.47	0.28	2.83
In	23.0	2.4	23.0	2.5	0.30	2.64	0.31	2.66
Pb	23.0	2.5	23.0	2.5	0.32	2.34	0.32	2.98
Ag	26.0	2.3	26.0	2.5	0.28	2.98	0.28	2.58
Cu	30.0	2.4	30.0	2.7	0.25	2.35	0.25	2.75
Au	26.0	2.4	26.0	2.5	0.28	2.68	0.28	2.77

The calculated results for $S(k)$ of these metals at their respective temperatures together with their experimental results entire momentum space are depicted in Figure 1.

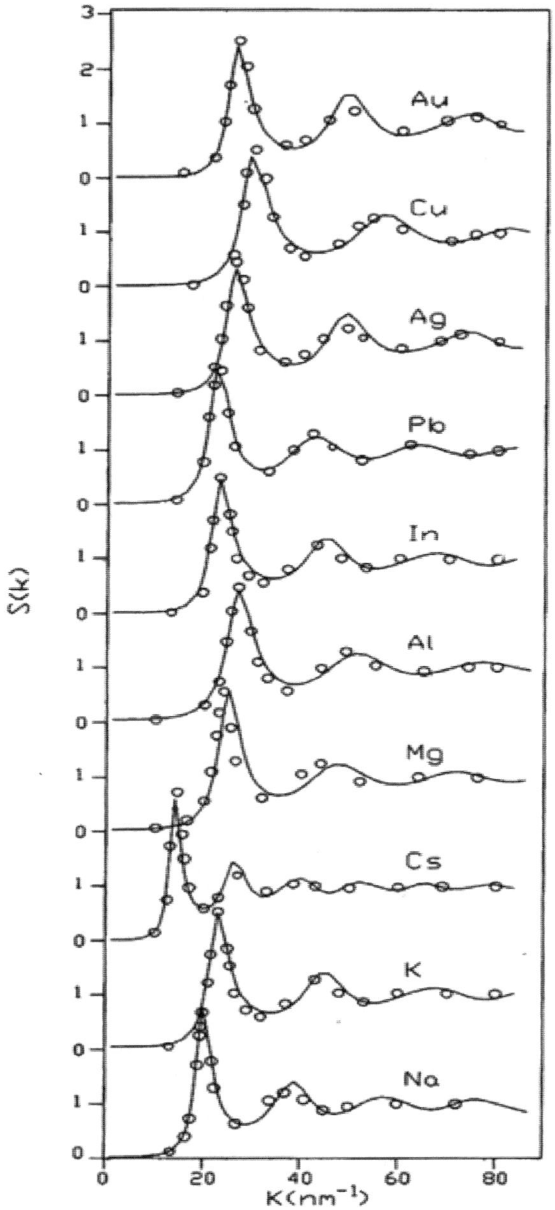

Figure 1. S(k) against k for liquid metals; (——) theoretical values, (o o o) experimental values.

We find from Figure 1 that structure factors of all the considered metals become constant around one in high k region. It shows the presence of short range order in liquid materials.

The peak positions and peak heights of the $g(r)$, computed from Eq. (11), for all the considered metals with their experimental values are listed in Table 2 and also presented in Figure 2(a) and Figure 2(b). The ratio of the positions of the first and second peaks of the calculated and the experimental $g(r)$ for all the liquids taken under investigation is about 0.51 and 0.53 respectively. This

Journal of Physics: Conference Series **765** (2016) 012012 doi:10.1088/1742-6596/765/1/012012

suggests that our model calculation of $g(r)$ gives the structural properties of real liquids fairly well. The deviation of $g(r)$ from unity is a measure of the local order around the reference atoms.

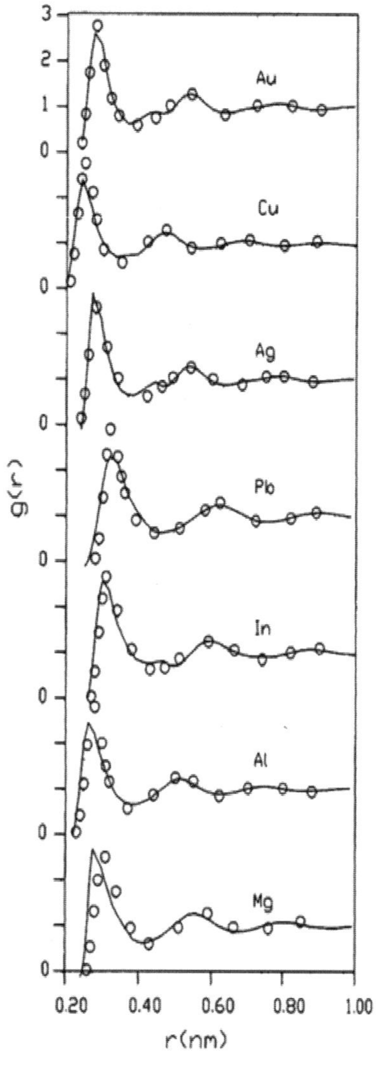

Figure 2(a). g(r) against r for liquid metals; (——) theoretical values, (o o o) experimental values.

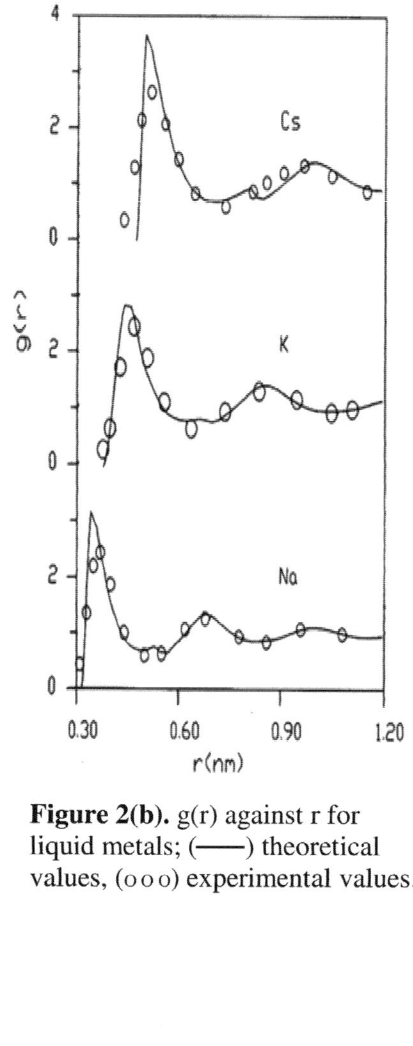

Figure 2(b). g(r) against r for liquid metals; (——) theoretical values, (o o o) experimental values.

Computed values of friction coefficients ξ_H, ξ_S and ξ_{HS} of the considered liquid metals are presented in Table 3. ξ_H is dominating in all cases as per our expectation because hard sphere repulsive forces dominating in the calculation of microscopic structural functions. However, ξ_S and ξ_{HS} also contribute significantly in the case of all liquid metals. These friction coefficients depend on soft part of the potential function.

TABLE 3. Friction coefficients of liquid metals due to hard sphere ξ_H, square well ξ_S and hard sphere- square well interaction ξ_{HS}.

Metals	$\xi_H \times 10^{-13}$(Kg/s)	$\xi_S \times 10^{-13}$(Kg/s)	$\xi_{HS} \times 10^{-13}$(Kg/s)
Na	7.41	1.56	1.94
K	7.55	1.12	1.58
Cs	7.56	1.50	1.89
Mg	11.78	1.36	1.19
Al	11.98	1.76	1.07
In	18.14	4.39	5.38
Pb	25.59	2.04	1.75
Ag	42.28	8.25	12.56
Cu	31.96	2.78	4.53
Au	62.19	11.05	18.65

The diffusivity in liquid metals has been evaluated under LT principle as mentioned in Ref. [4] using well known Einstein equation for coefficient of diffusion [29]. Computed results are compared with available experimental results [27] and presented in Table 4.

The surface tension is related with self-diffusion coefficients of liquids. Since self-diffusion coefficients of liquids are determined using microscopic structural functions along with SW potential and hence computed γ_{ST} is also related microscopic structure of liquids. Computed values of γ_{ST} with their experimental values taken from Ref. [23] are also presented in Table 4 and we find satisfactory agreement between them.

TABLE 4. Theoretical and experimental values of Diffusion coefficient, D(10^{-9} m^2/sec) and Surface tension, γ_{ST}(Nm^{-1}) of liquid metals.

Metals	$D_{Computed}$	$D_{Experimental}$	$\gamma_{STComputed}$	$\gamma_{STExperimental}$
Na	4.70	4.23	0.176	0.200, 0.197
K	4.28	3.76	0.096	0.110, 0.112
Cs	2.18	2.31	0.082	0.069, 0.070
Mg	5.23	5.63	0.418	0.557, 0.583
Al	4.93	4.87	0.801	0.867, 1.070
In	2.10	2.60	0.251	0.561
Pb	2.87	2.19	0.211	0.457, 0.462
Ag	2.61	2.55	1.026	0.925, 0.910
Cu	4.85	3.97	1.030	1.310, 1.320
Au	2.10	-	1.200	1.145, 1.138

Conclusions

In the present study, we present results of microscopic structural functions of ten liquid metals. The SW potential is analytically solved in repulsive and attractive regions and successfully applied for computing static structure factor and thermodynamic properties of the considered liquids, which are important information for metallurgical industry. Our results indicate that the SW model leads to a good agreement between computed and experimental results of structure factors and derived associated properties and hence such kind of theoretical works increase our confidence in present model. This model calculation provides an option to use the SW potential in the framework of MSMA to derive the various thermo-physical and thermodynamic properties of liquid metals without using any adjustable parameter.

Perturbation theory with hard sphere reference system is a good first approximation for the study of static and dynamic properties of liquid metals. The resulting correlation between D and γ_{ST} finds successful application for the estimation of γ_{ST}.

Acknowledgements

R K Mishra acknowledges the financial support from the Department of Science and Technology, Govt. of India, New Delhi under Fast Track Project (SR/FT/CS-021/2008) and R. Lalneihpuii to the University Grant commission, New Delhi for awarding Rajiv Gandhi National Fellowship.

References

[1] Gosh R C, Ziauddin Ahmed A Z and Bhuiyan G M 2007 *Eur. Phys. J.* B **56** 177

[2] Taylor M P, Strathmann J L and Lipson J E G 2001 *J Chem. Phys.* **114** 5654

[3] Gopala Rao R V and Venkatesh R 1989 *Phys. Rev. B* **39** 3563

[4] Venkatesh R and Mishra R K 2005 *J. Non-Cryst. Sol.* **351** 705

[5] Wertheim M S 1963 *Phys. Rev. Letts.* **10** 321

[6] Thiele E 1963 *J. Chem. Phys.* **39** 474

[7] Zalid F, Bhuiyan G M, Sultana S, Khaleque M A, Rashid R I M A and Rahman S M M 1999 *Phys. Stat. Sol.(b)* **215** 987

[8] Ravi S, Kalidoss M and Srinivasanmoorthy R 2001 *J. Amoros, Fluid Phase Equilibria.* **178** 33

[9] Prakash S G, Ravia R and Chhabra R P 2004 *Chem. Phys.* **302** 149

[10] Boulahbak M, Jakes N, Wax J F and Bretonnet J L 1998 *J. Chem. Phys.* **108** 2111

[11] Gopala Rao R V and Murthy A K 1974 *Phys. Stat. Sol. (b)* **66** 703

[12] Gopala Rao R V and Murthy A K 1975 *Phys. Letts.* **51A** 3

[13] Gopala Rao R V and Murthy A K 1975 *Z. NaturForsch* **30a** 619

[14] Mishra R K, Lalneihpuii R and Pathak R 2015 *Chem. Phys.* **457** 13

[15] Asherie N, Lomakin A and Bendek G B 1996 *Phys. Rev. let.* **77** 4832

[16] Noro M G and DiFrenkel 2000 *J. Chem. Phys.* **113** 2941

[17] Zaccarelli E, Foffi G, Dawson K A, Sciortino F and Tartaglia P 2001 *Phys. Rev. E.* **63** 031501.

[18] Cui J and Elliot J R 2001 *J. Chem. Phys.* **114** 7283

[19] Zhou Y, Karplus M, Ball K D and Berry R S 2002 *J. Chem. Phys.* **116** 2323

[20] Zhou Y, Karplus M, Wichert J M and Hall C K 1997 *J. Chem. Phys.* **107** 10691

[21] Blairs S 2006 *J. Colloid Interface Sc.* **302** 312

[22] Lebowitz J L and Percus J K 1966 *Phys. Rev.* **144** 251

[23] Lu H M and Jiang Q 2005 *J. Phys. Chem. B.* **109** 15463

[24] Block R, Suck J B, Freyland W, Hansel F and Glaser W 1977 *Proceedings of 3rd International Conference on Liquid Metals,* The Institute of Physics, Bristol, London 126

[25] Fowler R H 1937 *Proc. Roy. Soc A* **159** 229

[26] Born M and Green H S 1949 *A general kinetic theory of Liquids*, University Press, Cambridge

[27] Shimoji M and Itami T 1986 *Diffusion and Defect Data, Atomic Transport in Liquid Metals* Trans Tech Publication, Switzerland

[28] Waseda Y 1980 *The Structure of Non-crystalline Materials.* McGraw-Hill: New York

[29] Dahlborg U, Besser M, Kramer M J, Morris J R and Calvo-Dahlborg M 2013 *Physica B* **412** 50

Calculations of Photocurrent from BP and its band structures

Rebecca Lalngaihawmi* and R.K. Thapa
Condensed Matter Theory Research Group, Department of Physics, Mizoram University, 796004, India

rebeccalngi@gmail.com

Abstract. We present here the results of photocurrent from boron pnictides namely BP. The photocurrent in this system had been calculated by using three types of potential namely free-electron, Kronig-Penney and Muffin-Tin potentials. Schrodingers' equation has been solved by using these potentials to obtain the necessary initial and final state wave-functions for the electrons. These wave-functions are used for evaluating the matrix element $\langle\Psi_f|\boldsymbol{p}.\boldsymbol{A}+\boldsymbol{A}.\boldsymbol{p}|\Psi_i\rangle$ which is involved in the formula of photocurrent.

1. Introduction

Boron compounds have attracted research interest over the past few years as wide band gap semiconductors. For example, boron pnictides (BAs, BN and BP) are binary compounds belonging to III-V semiconductors, which crystallizes in zinc blende structure [1]. They are extensively under study and are highly attracted due to their high technological applications such as high-temperature, electronic and optical devices like linear and non-linear optics, solar cells, light-emitting diodes, etc. Photoemission studies from the surface of solids are mainly concerned with the excitation of electrons. These photoelectrons lying at the surface or at the bulk region needs sufficient energy to transport themselves across the barrier. However, the incident energy is too weak to provide the necessary momentum to excite the photoelectrons at the surface. Therefore, the spatial variations of vector potential at the surface are considered to be the main factor for the excitation of photoelectrons at the surface. For evaluation of the vector potential involved in the matrix element for transition from initial to final state, we require real and imaginary dielectric constants. These have been calculated by using the density functional theory based on full-potential approximation which is implemented in Wien2k code [2].

2. Methodology

The current density for photoemission is calculated by using the golden-rule approximations [3] as

$$\frac{dj}{d\Omega}=\frac{2\pi}{\hbar}\sum_f\left|\langle\Psi_f|H'|\Psi_i\rangle\right|^2\delta(E-E_f)\delta(E_f-E_i-\hbar\omega)f_0(E-\hbar\omega)[1-f_0(E)] \quad (1)$$

where Ψ_i and Ψ_f are the initial and final states wave-function and $H'=(e/2mc)(\boldsymbol{A}.\boldsymbol{p}+\boldsymbol{p}.\boldsymbol{A})$, $\boldsymbol{A}$ being the vector potential and $\boldsymbol{p}$ the one-electron momentum operator. In order to evaluate the matrix element in Eq. (1), we need to construct Ψ_i and Ψ_f and determine vector potential $\boldsymbol{A}$. Although the one-electron states are treated quite accurately in many photoemission calculations [4], the variation of the photon field $\boldsymbol{A}$ in the surface region is usually neglected. We have used a simple model for calculations of A as done by Bagchi and Kar [5].

We have used three types of potentials for calculating the wave function to evaluate the matrix element for calculating photocurrent. We will describe each one of these briefly here. First, the muffin-tin potential [6] had been used by which the initial state wavefunction derived is given by

$$\Psi_i(\boldsymbol{r}) = \begin{cases} \sum_L A_L f_{1l}(R) Y_L(\theta,\phi) & \textit{spherical region} \\ \sum_g \left[u_g e^{ik_g^+ .\boldsymbol{R}} + v_g e^{ik_g^+ .\boldsymbol{R}}\right] & \textit{interstitial region} \end{cases} \tag{2}$$

where $k_g^+ = \left[(\boldsymbol{k}+\boldsymbol{g})_\| \pm \sqrt{2(E_i - V_0) - |\boldsymbol{k}-\boldsymbol{g}|_\|^2}\right]$ and $\boldsymbol{R} = \boldsymbol{r} - \boldsymbol{C}_j$ with the $\boldsymbol{g}$'s denoting two-dimensional reciprocal lattice vectors and the $\boldsymbol{C}_j$is the origin at the jth layer.

The initial state wavefunction used for the free-electron model [7] is given by

$$\Psi_i(\boldsymbol{r}) = \begin{cases} \left[e^{ik_i z} + \frac{ik_i+\chi}{ik_i-\chi}\right] e^{i\boldsymbol{k}_\| \boldsymbol{r}_\|} & z \le 0 \\ \frac{2ik_i}{ik_i-\chi} e^{-\chi z} e^{i\boldsymbol{k}_\| \boldsymbol{r}_\|} & z > 0 \end{cases} \tag{3}$$

Where $\boldsymbol{k}_i^2 = \frac{2m}{\hbar^2} E_i - \boldsymbol{k}_\|^2$, $\chi^2 = \frac{2m}{\hbar^2}(V_0 - E_i) + \boldsymbol{k}_\|^2$, $\boldsymbol{k}_\|$ and $\boldsymbol{r}_\|$ are the components of $\boldsymbol{k}$ and $\boldsymbol{r}$ in the x-y plane i.e. the plane parallel to the surface.

Similarly for Kronig-Penney potential model, Schrodinger's equation had been solved for the bulk, surface and vacuum regions by wavefunction matching. The initial state wavefunction for this case is given by

$$\Psi_i(z) = \begin{cases} \left(1 - iPe^{-i\delta} sin\delta\right) - \left(P - ie^{i\delta} sin\delta\right) e^{-ik_i(z)} \\ Te^{-\chi z} \end{cases} \tag{4}$$

where $\chi^2 = \frac{2m}{\hbar^2}(V_0 - E_i)$ with V_0 being the step potential at the surface, P and T are reflection and transmission coefficients respectively. By wavefunction matching at the surface plane (z = 0), we find that

$$P = \frac{(\chi - ik_i) - (k_i - i\chi) e^{i\delta} sin\delta}{(\chi - ik_i) + (k_i - i\chi) e^{i\delta} sin\delta} \tag{5}$$

$$T = \frac{2ik_i sin2}{(\chi - ik_i) + (k_i - i\chi) e^{i\delta} sin\delta} \tag{6}$$

For all these three cases, the final state wave function [6] is a plane wave which is

$$\Psi_f(\boldsymbol{r}) = \begin{cases} \left[e^{ik_i z} + \frac{g-k_f}{g+k_f} e^{-gz}\right] & z \le 0 \\ \frac{2g}{g+k_f} e^{ik_f z} e^{i\boldsymbol{k}_\| \boldsymbol{r}_\|} & z > 0 \end{cases} \tag{7}$$

where $\boldsymbol{k}_f^2 = \frac{2mE_f}{\hbar^2} - k_\|^2$, $\boldsymbol{g}^2 = \frac{2m}{\hbar^2}\left(E_f - V_0\right) - k_\|^2$ and $E_f = E_i + \hbar\omega$.

3. Results and discussions

We have calculated photocurrent for normal photoemission from the surface and bulk states of BP as a function of photon energy. For our calculations, we have taken work function ϕ= 8.631 eV, potential energy V_0=15.95 eV, surface width a = 5.298 $\AA$, angle of incident radiation θ_i= $45^0$and scattering factor α=0.35. The initial and final state energies are taken to be E_i = 6.098 eV and E_F = 7.319 eV. For electromagnetic field calculation which is related to vector potential, we have used the real and imaginary parts of dielectric constants calculated by using the full-potential linearized augmented plane wave method implemented in Wien2k code [2] which is based on the density functional theory [8].The value of vector potential had been used as a subroutine in the main FORTRAN program to calculate photocurrent.

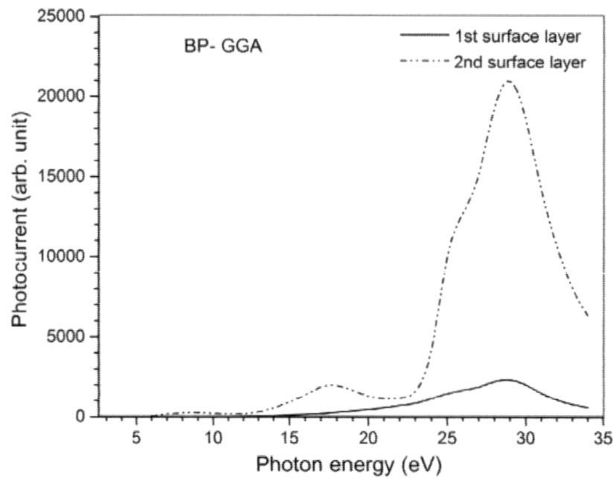

Figure 1. Plot of photocurrent in BP as a function of photon energy from the first two surface layers.

Plot of photocurrent as a function of photon energy ($\hbar\omega$) of the incident radiation for normal photoemission is shown in figure 1. In our calculations, the dielectric constants ε_1 and ε_2 were obtained using Wien2k code within generalized gradient approximation (GGA) [8]. The photocurrent increases with increase in photon energy. From the first surface layer, we found that there is small increase in photocurrent and a small bump at 29 eV photon energy. The trend of photocurrent from the second surface layer is different from that of photoemission from the first surface layer. Here, photocurrent shows a small peak at 18 eV and then decreases slightly after which it started increasing sharply giving a high peak at 30 eV. Beyond this, photocurrent then decreases with further increase in photon energy. The contribution of photocurrent from second surface layer showed similar trends as from the first layer, but occurrence of peaks are at different photon energies.

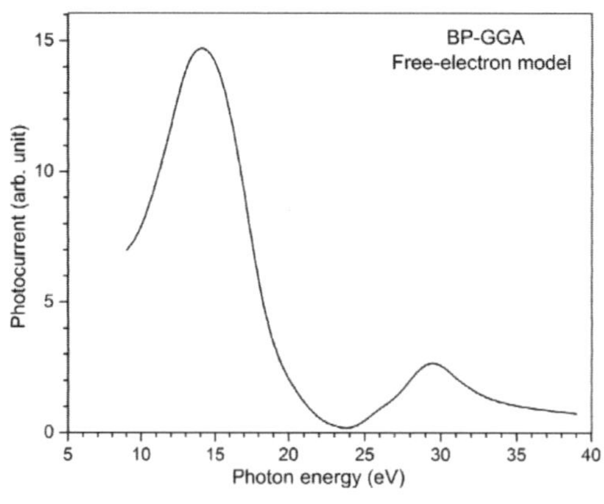

Figure 2. Plot of photocurrent in BP as function of photon energy using free-electron model.

In the case of free-electron model, photocurrent increases with increase in photon energy showing a small peak at 15 eV as shown in figure 2. Photocurrent then decreases and reaches a minimum at 23 eV which is known as the plasmon energy after which increase variably and reaches a maxima at 30 eV photon energy. Beyond this photon energy, the photocurrent decreases gradually.

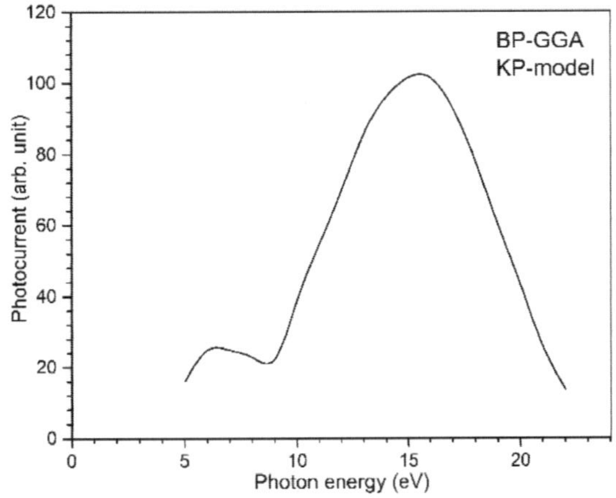

Figure 3. Plot of photocurrent as a function of photon energy using Kronig-Penney model in the case of BP.

In figure 3, we studied the variation of photocurrent as a function of photon energy using Kronig-Penney model and the results are obtained using the dielectric functions as already used in the surface layer calculations. We found here that the photocurrent shows a small peak at 7 eV and then decreases. As found earlier in the cases of other metals like aluminium, iron, tungsten etc., there is a minimum at 9 eV photon energy. The current rises above the plasmon energy and we a get another maximum at 16 eV of the incident radiation and decreases again. Thus, we can say that the surface variation of photon field is important in calculating the photocurrent cross-section also in the case of BAs. This fact has also been observed by Bagchi and Kar [5] in the case of tungsten. They have found that vector potential component in the surface region decreases greatly near the plasma frequency of the solids. The difference in the height of the peaks is due to change in the thickness of the surface.

Conclusions

The frequency dependence of the normal component of the photon field plays an important role in calculating the photoemission cross-section from the surface states of BP. The field component in the surface region normally increases with increase in photon energy and decreases greatly near the plasmon frequency of the metallic surface as seen in figures 1, 2 and 3 respectively. This explains the observed difference in normal photoemission from surface states near the plasmon frequency. We may thus state that the surface states play a very important role in the photoemission process.

Acknowledgement

RL is grateful to UGC (Delhi) for the NFST Fellowship, whereas RKT acknowledges a grant from SERB (DST, India) via grant No. EMR/2015/001407, Dt.11th March 2016.

References

[1] Mujica A, Angel Rubio A, Munoz A and Needs R J, "High pressure phases of group-IV, III-V and II-VI compounds", *Rev. Mod. Phys.*, vol. **75**, pp. 863-912, July 2003.

[2] Blaha, P, Schwarz S, Madsen M K H., Kvasnicka D, Luitz J, and Schwarz K, "An Augmented Plane Wave plus Local Orbitals Program for calculating Crystal Properties: *Wien2k User's Guide* (Wien: Technische Universtat Wien), 2015.

[3] Penn D R, "Photoemission spectroscopy in the presence of absorbate covered surface", *Phys. Rev. Letts.,* vol. **28**, pp. 1041-1044, October, 1972.

[4] Pendry J B, "Theory of Photoemission", *Surf. Sci.* vol. **57**, pp. 679-705, July 1976.

[5] Bagchi A and Kar N, "Refraction effects in angle-resolved photoemission from surface states on metals", *Phys. Rev. B,* vol. **18**, pp. 5240-5247, November 1978.

[6] Das P, and Kar N, "Photoemission calculations including both band structure effects and photon field variation" *Phys. Stat. Solidi b,* vol. **187**, pp. 551-556, February 1995.

[7] Thapa R K., and Kar N, "Photocurrent calculations in beryllium using a local dielectric model" *Phys. Rev. B,* vol. **5**, pp.17980-17983, June 1995.

[8] Perdew John P, Burke Kieron Burke and Ernzerhof M, "Generalized Gradient Approximation Made Simple", *Phys. Rev. Letts.,* vol. **77**, pp. 3865-3868, October 1996.

[9] Thapa R K Das P, and Kar N, "Frequency dependent photoemission from aluminium using a simple local dielectric model" *Physics Teacher,* vol. **33**, pp. 21-26, January 1991.

[10] Das P, Thapa R K, and Kar N, "Photoemission calculation with a simple model for the photon field: Application to Aluminium", *Modern Physics Letters B*, vol. **5**, pp. 65-72, January 1991.

[11] Thapa R K, and Davison S G, "Electromagnetic fields calculated in the surface region using a simple dielectric model and application to metals", *Indian Journal of Pure and Applied Physics,* vol. **34**, pp. 118-120, February 1996.

Calculation of photofield emission current by using the projection operator method of group theory

Gunakar Das[1]*, Lalrinpuii[1] Leishangthem Nirmala Devi[2] and Ram Kumar Thapa[2]

[1]Department of Physics, Government Kolasib College, Kolasib-796081, Mizoram, India.
[2]Department of Physics, Mizoram University, Tanhril, Aizawl-796004, Mizoram, India.

*Email: gunakardas@gmail.com

Abstract. A calculation of photofield emission is discussed in which initial state wave function has been deduced by using the projection operator method of group theory. A spatial dependent vector field is used to evaluate the matrix element for calculating the photofield emission current density.

1. Introduction

Photofield emission (PFE) is a technique in which a metal is irradiated by an incident laser radiation of photon energy. Photon energy is usually less than the work function of the metal under investigation. The incident radiation photoexcites the electrons to a final state which lies below the vacuum level, hence these electrons are confined within the metal surface. A strong static electric field of the order of 10^{11} V/m when applied to the surface of the metal causes the photoexcited electron to tunnel through the surface potential barrier into the vacuum region constitutes the photofield emission current (PFEC). In photofield emission, in addition to transmission probability *D(W)*, the transition matrix element $\langle \psi_f | A.p + p.A | \psi_i \rangle$ also plays an important role. In this matrix element, we find that the important ingredients are the vector potential A, initial state wave function ψ_i and final state wave function ψ_f There are several methods of deducing the initial state wavefunctionψ_i and the exactness of the model developed or used can correctly interpret the results in photofield emission calculations. For example, the LEED (low energy electron diffraction) states deduced by Pendry [1-2] for band structures and photoemission calculation are accurate, but the method is too complex.

One also uses the Green function method for obtaining the appropriate wavefunctions as has done by Bagchi [3] and Modinos [4]. Multiple scattering technique and density functional theory [5-7] are also being used in the interpretation of results of photofield emission.

The incident radiation in PFE can be used to probe the electron states between the Fermi level and the vacuum level. Therefore, the initial state of the electron can play an important role in determining the shape of the final state energy distribution. This allows one, therefore, to see the effect of initial state energy bands on PFE.

In this report, we are presenting calculation of photfield emission current by using the initial state wavefunction ψ_i which is deduced by projection operator method of group theory. As an example, basis function was derived for *W* (100) surface state. An appropriate dielectric model and vector potential *A* also will be used for the surface region.

2. Theoretical Formalism

2.1. Matrix element and transmission probability

We have considered a *p*-polarised radiation of photon energy to be incident on the metal surface. The surface normal is defined by the *z*-axis which is perpendicular to the *xy*-plane. The incident radiation is usually a laser beam, causes the transition of electrons from the initial state to final. We consider initial states to be electron states lying below the Fermi level, and final states are states in the vacuum (detector). Therefore, the photofield emission current density measured can be written [8] as

$$\frac{dj}{dE} = -\frac{e^3}{2\hbar^4\omega^3}\frac{n}{\Omega}(\hat{\varepsilon}.\hat{z})^2 f(E-\hbar\omega)\int_{-V_o+\hbar\omega}^{E} dW \frac{D(W)\left|M_{fi}\right|^2}{\left[W(W-\hbar\omega)\right]^{\frac{1}{2}}} \tag{1}$$

Where $\frac{n}{\Omega}(\hat{\varepsilon}.\hat{z})^2 = \left|\frac{A_\omega(z)}{A_0}\right|^2 = \left|\tilde{A}_\omega(z)\right|^2$ and $A_\omega^z(z)$ is the *z*-component of vector potential along the *z*-axis, A_0 is the amplitude of vector potential associated with the incident radiation. Also $f(E-\hbar\omega)$ is the Fermi-Dirac distribution function. The matrix element M_{fi}, when expanded in one dimension along z-axis is given by,

$$\begin{aligned} M_{fi} = &\int_{-d}^{0}\psi_f^* A_z \frac{dV}{dz}\psi_i dz + \int_{-d}^{0}\psi_f^* \frac{d^2A_z}{dz^2}\left(-i\eta\frac{d}{dz}\right)\psi_i dz + \\ &\int_{-d}^{0}\psi_f^* \frac{dA_z}{dz}\left(-\eta^2\frac{d^2}{dz^2}\right)\psi_i dz + \int_{-d}^{0}\psi_f^* \frac{dA_z}{dz}\psi_i dz. \end{aligned} \tag{2}$$

The transmission probability *D(W)* in Eq. (1) describes the quantum mechanical transmission that the photoexcited electrons with energy

$$E = W + \frac{\hbar^2.k_{\|}^2}{2m} \tag{3}$$

will travel across the surface potential barrier which is deformed by the applied electrostatic field and the image potential barrier. Here *W* is the normal component of energy *E*. *D(W)* in Eq. (1) is obtained by solving [9]Airy's differential equation and is given by

$$D(W) = \frac{W^{\frac{1}{4}}\sqrt{\pi}}{(\eta eF)^{\frac{1}{6}}}\left(\frac{2ik_i}{ik_i+\chi}\right)(2m)^{\frac{1}{12}}\exp\left[-i\left(\frac{2}{3}\frac{W^{\frac{3}{2}}\sqrt{2m}}{\eta eF}+\frac{\pi}{4}\right)\right] \tag{4}$$

2.2. Dielectric Model and Vector Potential

In photofield emission, the dominant contribution to the photocurrent comes from the surface photo effect. As evidenced by several authors, for example, Levinson *et. al.*, [10] Fiebelman [11], Kliewer [12], the main contribution to the surface photoeffect is due to spatial variation of photon field vector

A. A systematic calculation of this effect on surface photoeffect had been also done by Thapa [13] *et.al.* It is, therefore, necessary that a detailed theory of dielectric model is developed at first for the appropriate formalism of ***A***. In this formalism here, we will employ the dielectric model of Bagchi and Kar [14], which takes into account the dependence of ***A*** not only on frequency ω but also on z-coordinate. Therefore, the vector potential for the surface region $(-d \le z \le 0)$ in one dimension is given by,

$$A_z(\omega, z) = -\frac{A_o d\varepsilon(\omega)}{[1-\varepsilon(\omega)]z + d} \tag{5}$$

where .

$$A_o = \frac{\sin 2\theta_i}{\left[\varepsilon(\omega) - \sin^2\theta_i\right]^{\frac{1}{2}} + \varepsilon(\omega)\cos\theta_i}$$

In the above equation, θ_i is the angle of incident photon radiation with respect to the surface normal.

3. Description of Initial and Final state wave Functions

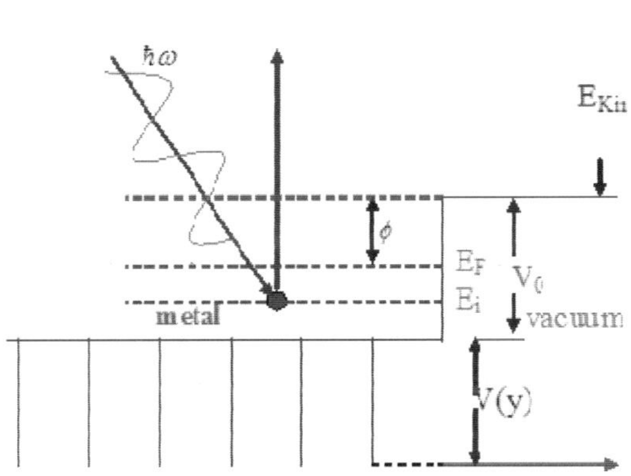

Figure 1: Schematic representation of Kronig-Penney δ-potential model for calculating the initial state wave function by using projection operator method of Group Theory.

To calculate the initial state wavefunction ψ_i, we have assumed the crystal potential of the solid which is defined by ∂- potential and represented by the Kronig-Penny potential. The potential is periodic with the periodicity of the lattice as shown in Fig. In one dimension, one can write ψ_i as

$$\psi_i(z) = \begin{cases} \psi(z) + R\psi^*(z) & z \le 0 \; (bulk \,\&\, surface) \\ T e^{-\chi z} & z \ge 0 \; (vacuum) \end{cases} \tag{6}$$

where $\psi^*(z)$ is the complex conjugate of $\psi(z)$, R is the reflection coefficient, T is the transmission coefficient across the boundary plane and $\chi_0^2 = \frac{2m}{\hbar^2}(-V_0 - E_i)$ with V_0 as the potential at the surface which an electron encounters while transmitting through the boundary surface. Matching the wavefunction and its derivatives at $z = 0$ gives the value of coefficients R and T as

$$R = \frac{-\chi + \mu - ik}{\chi - \mu - ik} \tag{7}$$

and $$T = \sigma\left(\frac{-2ik}{-\chi - \mu - ik}\right) \tag{8}$$

where $$\sigma = \frac{p}{k_i a}\psi(0)\frac{\sin k_i a}{\cos ka - \cos k_i a} \tag{9}$$

and $$\psi(0) = -2iC\frac{\sin k_i a}{1 + \cos k_i a - i\sin k_i a} \tag{10}$$

Also, p is the strength of the ∂- potential barrier and it is assumed to be positive and μ is used as converging factor.

For *W*(100) surface state occurs [15] in the energy band gap $\left(\Delta_2 - \Delta_2'\right)$. The linear combination of atomic orbital (LCAO) representation for Δ_2' is $\frac{1}{3}\left(3z^2 - r^2\right)$, where $r^2 = x^2 + y^2 + z^2$. We have, therefore, considered the point Δ_2' for which the point group is C_{4v}. We have obtained the basis function for the C_{4v} point group corresponding to *W* (100) surface state by using projection operator formula [16]

$$\mathrm{P}^{p}_{mn} = \frac{l_p}{g}\sum_T \Gamma^p(T)^*_{mn} P(T) \tag{11}$$

Here l_p is the dimension of the unitary irreducible representation of the group G, *g* is the order of G and $\sum_T$ is the summation over all the transformation *T* of G.

Now introducing the atomic orbital $\Phi(z)$, which includes the basis function derived by projection operator method of group theory obtained from Eq. (11), the final form of initial state wavefunction can be represented by

$$\psi_i(z) = \begin{cases} \sigma\left[\Phi(z)e^{ik\,z} + R\Phi^*(z)e^{-ik\,z}\right], & bulk\ \&\ surface\, z \le 0 \\ T e^{-\chi z}, & vacuum\ z \ge 0 \end{cases} \tag{12}$$

The initial state wavefunction given by Eq. (12) has been used in conjunction with final state wavefunction ψ_f given by Eq. (5) and vector potential $A_\omega(z)$ of Eq. (3), to calculate photofield emission current by using the formula given in Eq. (1).
For calculation of *PFEC* with the inclusion of ψ_i defined in Eq. (12) are evaluated by writing FORTRAN programme.
The description of the final state wavefunction $\left|\psi_f\right\rangle$ used is already given elsewhere [17] as

$$\psi_f(z)=\begin{cases}\left(\dfrac{m}{2\pi\eta^2 q_f}\right)^{\frac{1}{2}}\dfrac{2q_f}{q_f+k_f}e^{-\alpha|z|}e^{ik_f z}.e^{i\vec{k}_{\|}.\vec{R}_{\|}}, & z<0\ \ (bulk\ \&\ surface)\\ \left(\dfrac{m}{2\pi\eta^2 q_f}\right)^{\frac{1}{2}}\left[e^{iqz}+\dfrac{q_f-k_f}{q_f+k_f}e^{-iq_f z}\right].e^{i\vec{k}_{\|}.\vec{R}_{\|}}, & z>0\ \ (vacuum)\end{cases} \tag{13}$$

Where, $k_f^2=\dfrac{2mE_f}{\eta^2}-\vec{k}_{\|}^2\ ;\ q_f^2=\dfrac{2m}{\eta^2}\left(E_f+V_0\right)-\vec{k}_{\|}^2\quad and\ \ E_f=E_i+\eta\omega$

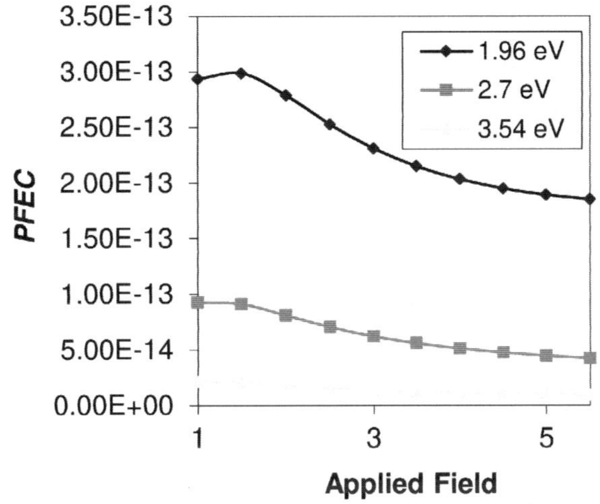

Figure.2 Variation of *PFEC* against the applied field in the case of *W*.

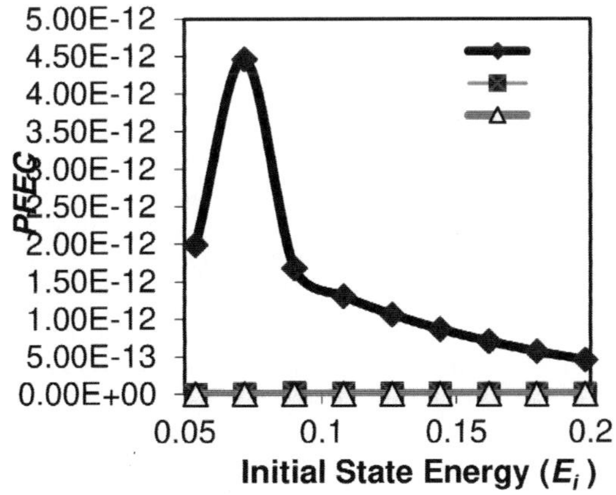

Figure 3. Variation of photofield emission current (*PFEC*) against initial energies of the electron below the Fermi level.

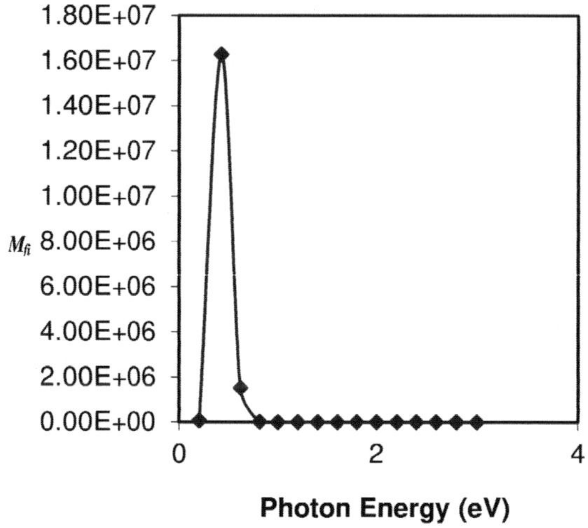

Figure. 4 Variation of matrix element M_{fi} against photon energy ($\eta\omega$).

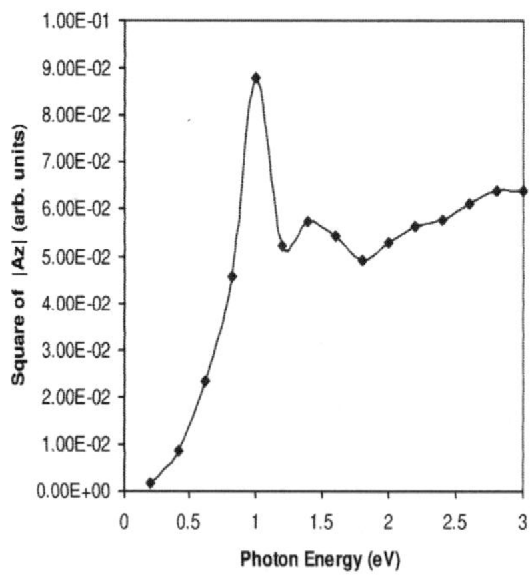

Figure 5. Plot of variation of vector potential $|A_z|^2$ against photon energy ($\eta\omega$) .

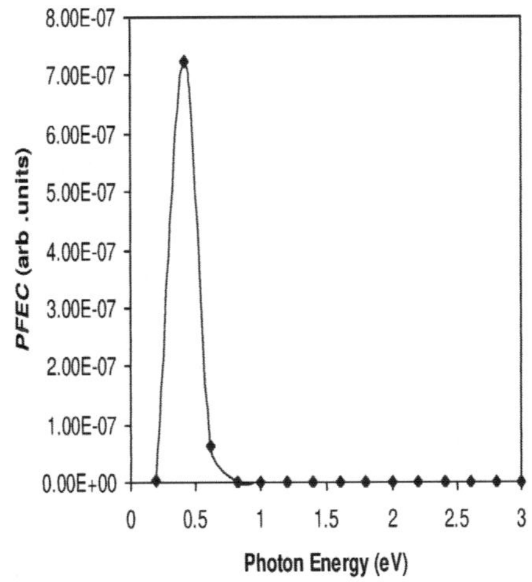

Figure 6. Variation of photofield emission current (*PFEC*) against photon energy ($\eta\omega$).

4. Results and Discussions

We discuss here the photofield emission current (*PFEC*) in the case of metal *W* as a function of initial state energy (E_i), photon energy ($h\omega$) and the applied high static electric field (*F*). We have also plotted the variation of $|A_z|^2$ against photon energy ($h\omega$).

In Fig. 2, we show the plot of variation of photofield emission current (*PFEC*) as a function of applied field forE_i = 2.1488 *eV*, $\theta_i = 45^0$ for three values of photon energies 1.96 *eV*, 2.70 *eV* and 3.54 *eV*. Here also, we find that the magnitude of *PFEC* is higher for low photon energies and decreases as photon energies increase. However, the variation in *PFEC* as applied field increases shows similar trends for all the three cases of photon energies. The decrease in *PFEC* is found to be exponential with the further decrease in the applied field.

We have also shown in Fig. 3, the variation of *PFEC* as a function of initial state energy of the electron with respect to Fermi level (E_F) as a reference level. In this case, also, *PFEC* is plotted for the same three photon energies $\eta\omega = 1.96$ *eV*, 2.70 eV, and 3.54 *eV*, but for values of applied field F = 3.08 X 10^{11} *V/m* and $\theta_i = 45^0$. In this case also, for low photon energies $\eta\omega = 1.96$ *eV*, the variation of *PFEC* is interesting. It showed a maximum in *PFEC* at E_i = 0.75 *eV* below E_F, and decreases with the value of E_i located further away from E_F. For other values of photon energies, we find that the values of *PFEC* are very low in magnitude remaining constant without exhibiting any peaks.

Fig. 4 shows the plots of variation of matrix element M_{fi} against photon energy ($\eta\omega$) for the value of applied field *F* = 3.08 X 10^{11} *V/m*, E_i = 2.1488 *eV* below $E_F$and $\theta_i = 45^0$. It is interesting to note here that as the values of photon energy of the incident radiation increases for low value, M_{fi} increases and showed a maximum at $\eta\omega$ = 0.50 *eV*. Beyond $\eta\omega$ = 0.50 *eV* , M_{fi} decreases and with

the further increase in $\eta\omega$, it constantly showed a minima value. We have plotted M_{fi} only within very low photon energy values for $\eta\omega$ = 0 to 3 *eV*.
The occurrence of origin of the peak in M_{fi} as shown in Fig. 4 can be understood from the plot of $|A_z|^2$ against $\eta\omega$ as shown in Fig. 5. We find that $|A_z|^2$ slowly increases as $\eta\omega$ is increased and showed a maximum at $\eta\omega$ = 1.0 *eV*. With the further increase in $\eta\omega$ beyond 1 *eV*, the value of $|A_z|^2$ decreased and a dip was shown at $\eta\omega$ ~ 1.25 *eV*. The second hump of small height in $|A_z|^2$ was observed at $\eta\omega$ ~ 1.5 *eV*. We can attribute the occurrence of maxima in M_{fi} at $\eta\omega$ = 0.5 *eV* due to the occurrence of maxima in $|A_z|^2$ at $\eta\omega$ = 0.5 *eV*, although the shift in $\eta\omega$ is about 0.5 *eV* in the occurrence of maxima in M_{fi} and $|A_z|^2$.
In Fig. 6, the plot of *PFEC* as a function of photon energy is shown for values of applied field F = 3.08 X 10^{11} *V/m*, $E_i = 2.1488$ *eV* and $\theta_i = 45^0$. We find that *PFEC* showed an exponential decrease with the decrease in applied field (Fig 2). It had shown high order magnitude in *PFEC* for low photon energies which is due to the occurrence of photon frequency in the formula of *PFEC* (Eq.1) where it varied inversely with frequency. The exponential decrease in *PFEC* is due to the presence of exponential term which is a predominant factor in photofield emission. This fact had been already seen also in the case of field emission [15]. However, in field emission due to dominant characteristics of the exponential term, usually a term called *R* factor is plotted which in fact nullifies the effect of an exponential factor to field emission current calculated.
We have also seen the effect of the variation of $|A_z|^2$ as a function of photon energies which in fact contribute to the values of the matrix element (M_{fi}). Thus, we find that matrix element M_{fi} is not a constant factor when checked with the variation of $\eta\omega$. The change in M_{fi} in fact contributes to the photofield emission current. This is contradicting the report of Lee and Schaich[19] where it was observed that M_{fi} when plotted against $\eta\omega$ remained constant. From our computations, we find therefore that calculation of the matrix element M_{fi} is equally important in photofield emission and its evaluation must be considered in details.
Just like in the case of free electron model, we find that the plot of photofield emission current as a function of initial state energy showed similar features. The peak height in *PFEC* is too low for higher photon energies. This implied that only at low photon energy, photofield effect is prominent.

5. Results and Discussions

Gao and Reifenberger [8] have given a detailed study of photofield emission from *W* but the theoretical interpretation was based on free electron concept. Keeping in mind that *W* is a strongly bonded metal and application of free electron model may not be appropriate, we have presented here a model for photofield emission calculation. This model has also been used by Thapa *et al* [18]in photoemission calculations and the results on metal and semiconductors showed qualitative agreement with the experimental results. However, the point of difference with the model presented here is the dielectric model for the calculation of vector potential. With this vector potential, we intend to calculate photofield emission current in the strongly bonded metals like *W*, *Mo* etc. and compare the results with the other methods of calculations and the experimental values as given by Gao *et al*[8].

Acknowledgment
GD acknowledges the financial support from University Grants Commission, New Delhi, India under the Grant No. F.No. 42-822/2013 (SR)

References

[1] Pendry JB 1974. Low Energy Electron Diffraction, *Academic Press, London* pp 153-220.
[2] Pendry JB 1976. Theory of photoemission, *Surf. Sci.***57**, 679-705.
[3] BagchiA 1974. Theory of surface effect in photo-assisted field emission, *Phys. Rev.***B 10**, 542-553.
[4] Modinos A 1976.A green function technique for calculating total energy distributions of field-emitted electrons, *J. Phys. **C** : Solid State Phys.* **9**, 3867-3876.
[5] Ernst A,HenkJ andThapaRK 2005. Ultrathin antiferromagnetic films on a ferromagnetic substrate: a first principles study of Mn on Fe(001), *J. Phys. Condens. Mat.***17**, 3269-3283.
[6] Durham PJ 1981.Theory of photoemission from random alloys, *J. Phys.***F 11**, 2475-2493.
[7] Haggmann Mark J 1997. Simulations of laser-assisted field emission within the local density approximation of Kohn-Sham density-functional theory, *Intl. J. Quantum Chemistry***65**, 857-865.
[8] Gao Y 1986.*Photofield Emission Study of Metal Surfaces* Ph.D. thesis, Purdue University, Purdue.
[9] DasG2005. *A Theoretical study of Photofield emission from metals incorporating the spatially dependent Photon fields* ,Ph.D. Thesis, Assam University, Silchar.
[10] Levinson HJ, Plummer EW and FeibelmanPJ 1979. Effects on the photoemission of the spatially varying photon field at a metal surface, *Phys. Rev. Lett.* **43**, 952-955.
[11] Feibelman PJ 1975. Self-consistent calculation of the surface photoelectric effect, *Phys. Rev. Lett.***34**, 1092-1095.
[12] KL Kliewer1977. Surface photo effect for metals: Energy and angular electron distributions, *Phys. Rev.* **B 15**, 3759-3774.
[13] Thapa RK and Kar N 1990. Electromagnetic fields near surfaces in a simple model: the case of aluminium, *Indian J. Phys.***64A**, 321-324; Das P, Thapa RK and Kar N 1991. Photoemission calculation with a simple model for the photon field: Application to aluminium, *Mod. Phys. Lett.***B 5**, 65-72.
[14] Bagchi A and Kar N 1978. Refraction effects in angle-resolved photoemission from surface states on metals, *Phys. Rev.* **B 18**, 5240-5247.
[15] Kar N 1976. *A Theoretical Study of Effects of Surface States and Resonances on Field Emission* (Ph.D. Thesis, University of Pennsylvania).
[16] Cornwell JF 1969. *Group Theory and Electronic Energy Band in Solids* (North-Holland Publication), pp 54.
[17] Thapa RK and r Das Gunaka 2005. "A simple theory of photofield emission from the surface of metal", *Intl. Jour. Mod. Phys.* **B 19**, 3141-3149.
[18] Zoliana B, Pachuau Z , Zadeng L, Patra PK, Khating DT, Thapa2003 RK. A model photoemission calculations using projection operator method, *Intl. J. Mod. Phys.***B 17**, 2897-2902.
[19] Lee J.T. and Schaich W.L 1988, *Phys.Rev.***B 38**, 3747.

Journal of Physics: Conference Series **765** (2016) 012015 doi:10.1088/1742-6596/765/1/012015

A theoretical study of Photofield emission in Gallium Arsenide using Kronig-Penney potential model

Leishingthem Nirmala Devi[1] and Ram Kumar Thapa[2]

[1]Condensed Matter Theory Research Group, Physics Department, Mizoram University Aizawl 7896 004, Mizoram India.
[2]Condensed Matter Theory Research Group, Physics Department, Mizoram University Aizawl 7896004, Mizoram India.

nirmalairom18@gmail.com

Abstract. In this report, we have shown the variation of photofield emission current as a function of applied electric field, the photon energy and initial state energy of the electrons with reference to the Fermi level. The photofield emission current is calculated with the help of Kronig- Penney model potential. The origin of peak in photofield emission current in the valence band is explained with the help of result of density of state calculated.

1. Introduction

In this report, we are presenting the calculated results of photofield emission current for gallium arsenide (GaAs) by using Kronig-Penney potential model [1]. PFEC is calculated by using the formula given by Gao and Reifenberger [2]. For the calculation of PFEC, the initial state wavefunction as deduced by Thapa and Kar [1] by using the Kronig-Penney potential model will be used. Dielectric model deduced by Bagchi and Kar [3] is used for obtaining the vector potential. The calculated results of PFEC are plotted as a function photon energy, applied electric field and initial state energy. The maximum peak occurs in the plot of PFEC against initial state energy with respect to Fermi energy is correlated with the plot of density of state (DOS) of GaAs obtained by using density functional theory (DFT) which is implemented in wien2k code [4].

2. Theoretical Formalism

A p-polarised radiation of photon energy $\hbar\omega$ to be incident on the metal surface. This incident radiation, usually a laser beam, causes the transition of electrons from the initial state ψ_i to final state ψ_f. We consider initial states to be electron states lying below the Fermi level, and final states are states in the vacuum (detector). Therefore, the photofield emission current density measured can be written Gao and Reifenberger as [2].

$$\frac{dj}{dE} = -\frac{e^3}{2\hbar^4\omega^3}\frac{n}{\Omega}(\hat{\varepsilon}.\hat{z})^2 f(E-\hbar\omega)\int_{-V_o+\hbar\omega}^{E} dW \frac{D(W)\left|M_{fi}\right|^2}{\left[W(W-\hbar\omega)\right]^{\frac{1}{2}}} \quad (1)$$

Where $\frac{n}{\Omega}(\hat{\varepsilon}.\hat{z})^2 = \left|\frac{A_\omega(z)}{A_0}\right|^2 = \left|A_\omega^o(z)\right|^2$, and $A_\omega^z(z)$ is the *z*-component of vector potential along *z*-axis, A_0 is the amplitude of vector potential associated with the incident radiation, $f(E-\hbar\omega)$ is the Fermi-Dirac distribution function, *D(W)* is the quantum mechanical transmission probability and the energy of the photo-excited electrons is $E_{KIN} = W + \frac{\hbar^2 k_\parallel^2}{2m}$, where *W* is the normal component of kinetic energy E_{KIN}. These electrons will travel across the surface potential barrier which is deformed by the applied electrostatic field and the image potential barrier. At first we will discuss the various parameters involve in Eq. (1).

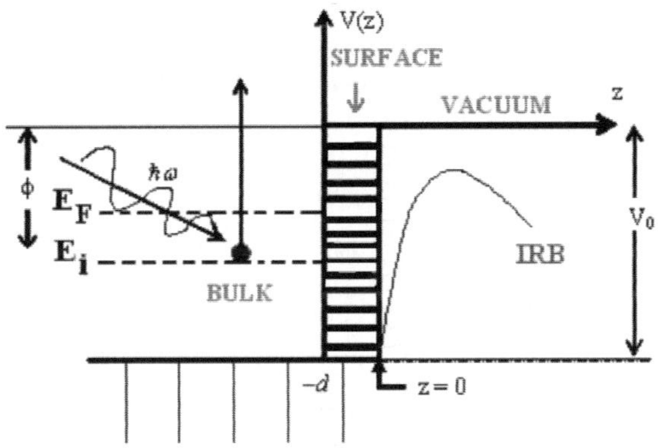

Figure 1. Model potential used for photofield emission calculation.
Here E_i is the Fermi level E_i is the initial state energy of the electron,
IRB is the image rounded barrier potential, V_0 is the potential barrier.

Matrix element M_{fi} in Eq. (1) due to transition of electrons from the initial state $|\psi_i>$ to the final state $|\psi_f>$ is

$$M_{fi} = \langle f|\boldsymbol{A}.\boldsymbol{p} + \boldsymbol{p}.\boldsymbol{A}|i\rangle \tag{2}$$

Now for Hamiltonian, $\boldsymbol{H} = \frac{\boldsymbol{p}^2}{2m} + V$ where momentum operator $p = -i\hbar\Delta$, the commutator ***A.p*** with the Hamiltonian ***H*** is given by,

$$[\boldsymbol{A.p}, \boldsymbol{H}] = -i\hbar \boldsymbol{A}.\nabla V + \frac{\hbar^2}{2m}\nabla^2(\boldsymbol{A}.\boldsymbol{p}) + \frac{i\hbar}{m}\nabla(\boldsymbol{A}.\boldsymbol{p}).\boldsymbol{p} \tag{3}$$

From Eq. (2) and Eq. (3), we have

$$M_{fi} = \frac{2i}{\omega}\langle f|\boldsymbol{A}.\nabla V|i\rangle + \frac{\hbar}{m\omega}\langle f|\nabla^2\boldsymbol{A}.\boldsymbol{p}|i\rangle - \frac{2i}{\omega}\langle f|\nabla.(\boldsymbol{A.p}).\boldsymbol{p}|i\rangle - i\hbar\langle f|\nabla.\boldsymbol{A}|i\rangle \tag{4}$$

In Eq. (4), $V = V_B + V_S$, where V_B and V_S are the bulk and surface potentials. In photofield emission, the dominant contribution to the current comes from the surface photo effect. In Eq. (4), the effective potential is given by

$$V(z) = \begin{cases} -V_0 - eFz - \frac{e^2}{4z}, & for\ z > 0 \\ 0 \quad , & for\ z < 0 \end{cases} \tag{5}$$

where $V(z)$ is called image rounded potential barrier.

To evaluate the Initial state wave functions $\psi_i(z)$, one solves the one-dimensional Schrodinger equation

$$\frac{d^2\psi(z)}{dz^2} + k_i^2\,\psi(z) = -2V(z)\,\psi(z) \tag{6}$$

where $k_i^2 = 2E_i$ and $V(z)$ is the δ-function potential of the Kronig-Penney (K-P) model.

Let $\phi(z)$ denote the Bloch wave function deep inside the metal and $\phi^*(z)$ the time version of $\phi(z)$. The eigenfunction in the semi-infinite solid $(z<0)$ was chosen to have the form

$$\psi_i(z) = \phi(z) - P\phi^*(z) \tag{7}$$

Where P is the reflection coefficient obtained by matching the wave function and its derivative at z = 0. The initial state wavefunction for z < 0 may be written as

$$\psi_i(z) = \left(1 - iP\,e^{-i\delta}\sin\delta\right)e^{ik_i z} - \left(P - ie^{-i\delta}\sin\delta\right)e^{-ikz} \tag{8}$$

Where $\cot\delta = \frac{k_i}{g}$, g being the strength of the potential. The Initial state wave functions outside the metal (z > 0) is

$$\psi_i(z) = Te^{-\chi z} \tag{9}$$

Where T being the transmission coefficient across the boundary plane and

$$\chi^2 = 2(V_0 - E_i) \tag{10}$$

Where V_0 is the surface step potential, the matching boundary conditions at z = 0 is given as

$$P = \frac{(\chi - ik_i) - (k_i - i\chi)\,e^{i\delta}\sin\delta}{(\chi - ik_i) + (k_i - i\chi)\,e^{-i\delta}\sin\delta} \tag{11}$$

$$T = \frac{2k_i\sin 2\delta}{(\chi - ik_i) + (k_i - i\chi)\,e^{-i\delta}\sin\delta} \tag{12}$$

and the final state wave functions $\psi_i(z)$ is given by

$$\psi_f(z) = \begin{cases} \left(\frac{1}{2\pi q}\right)^{\frac{1}{2}} \frac{2q}{q+k_f} e^{ik_f z}, & z<0\,(bulk\ \&\ surface) \\ \left(\frac{1}{2\pi q}\right)^{\frac{1}{2}} \left[e^{iqz} + \left(\frac{q-k_f}{q+k_f}\right)\right] e^{-iqz}, & z>0\,(vacuum) \end{cases} \tag{13}$$

where, $k_f^{\,2} = 2E_i$, $q^2 = 2(E_f - V_0)$ *and* $E_f = E_i + \hbar\omega$.

The matrix element M_{fi} in Eq. (4) when expanded in one-dimension along z-axis is given by,

$$M_{fi} = \int_{-d}^{0} \psi_f^* A_z \frac{dV}{dz}\psi_i dz + \int_{-d}^{0} \psi_f^* \frac{d^2 A_z}{dz}\left(-i\hbar\frac{d}{dz}\right)\psi_i dz + \int_{-d}^{0} \psi_f^* \frac{dA_z}{dz}\left(-\hbar^2\frac{d^2}{dz^2}\right)\psi_i dz + \int_{-d}^{0} \psi_f^* \frac{dA_z}{dz}\psi_i dz \tag{14}$$

Here A_z is the z-component of vector potential along z-axis. A_z will be the one used by Thapa and Das [5] for calculating PFEC in the case of W.

The transmission tunneling probability *D(W)* used in Eq. (1) has been calculated Thapa *et al.*[5] by solving the Airy's differential equation and matching the wave functions inside and outside the surface at z = 0.0. The standard form of Airy's differential equation is

$$D(W)=\frac{W^{\frac{1}{4}}\sqrt{\pi}}{(\mathrm{h}eF)^{\frac{1}{6}}}\left(\frac{2ik_i}{ik_i+\chi}\right)(2m)^{\frac{1}{12}}\exp\left[-i\left(\frac{2}{3}\frac{W^{\frac{3}{2}}\sqrt{2m}}{\mathrm{h}eF}+\frac{\pi}{4}\right)\right] \quad (15)$$

The wavefunction given by Eq. (8) and Eq. (13) are then used in Eq. (1) to calculate PFEC by writing FORTRAN program.

3. Results and Discussions

We report here the calculated results of photofield emission current (PFEC) of gallium arsenide (GaAs) obtained by using the Kronig-Penney potential model. PFEC is calculated as a function of initial state energy (E_i), photon energy ($\mathrm{h}\omega$) and the applied high static electric field (F) at surface and bulk region. The input parameters used for calculations of PFEC are: surface width (d) = 5.7363 A^0, initial state energy (E_i) = 4.4988 eV, potential barrier height (V_0) = 10.2562 eV, work function (ϕ) = 4.69 eV, Fermi energy (E_F) = 5.5662 eV, scattering factor (α) = 0.35 and phase shift (δ) = -0.6416.

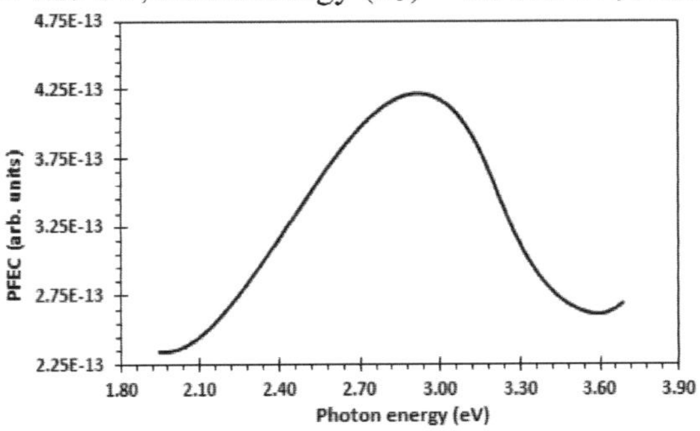

Figure 2. Plot of variation of PFEC against the photon energy. The angle of incident photon radiation is $\theta_i = 45^0$ and surface width (d) = 5.7363A^0.

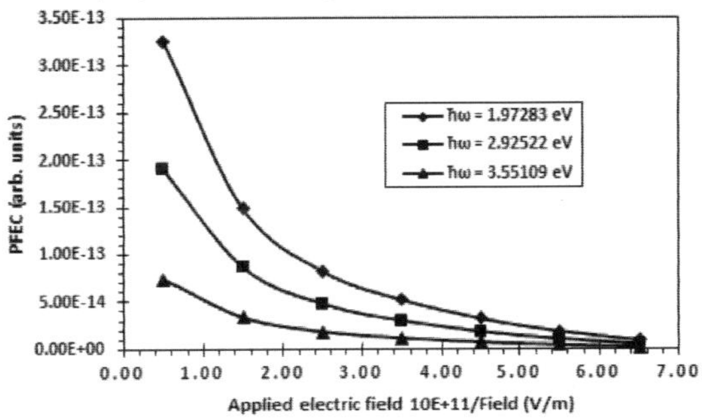

Figure 3. Plot of variation of PFEC against the applied electric field for three different values of photon energies and the angle of incident photon radiation is $\theta_i = 45^0$.

In Fig. 2, we have plotted the calculated results of calculated PFEC for GaAs as a function of photon energy $\mathrm{h}\omega$. In this case, we have taken applied electric field (F) is equal to 1.6 x 10^{11} V/m and calculated PFEC for photon energy in a range of $\mathrm{h}\omega$ = 1.80 – 3.90 eV. Here we have chosen the initial state energy E_i = 1 eV below Fermi level (E_F = 0.0). From the plot we have seen that the PFEC is increased as increase of photon energy. It shows a maximum at photon energy 2.925 eV. With the further increase in photon energy PFEC decreases and it becomes minimum at around $\mathrm{h}\omega$ = 3.60 eV.

In Fig. 3, we have plotted the calculated results of PFEC as a function of the applied electric field (F) for three difference values of photon energies $\mathrm{h}\omega$ = 1.97283 eV, 2.92522 eV and 3.55109 eV respectively. We have chosen the initial state energy E_i = 1 eV below Fermi level (E_F = 0.0). From the plot, we have seen that the value of applied electric field when increase the PFEC decreased from a high value towards minimum in an exponential manner for all the three different values of photon energies. The exponential decrease in PFEC is due to reason that an exponential term is present in calculation of transition probability $D(W)$ given by Eq. (15). From this plot we have also seen that, the value of PFEC maximum for low photon energy $\mathrm{h}\omega$ = 1.97283 eV and the value of PFEC minimum for large photon energy $\mathrm{h}\omega$ = 3.55109 eV. This is due to the reason that PFEC given in Eq. (1) is inversely proportional to cube of frequency of incident photon radiation. However, the variation of PFEC in all the cases of three different values of photon energies are similar in nature.

In Fig. 4, we have plotted the results of the calculated PFEC as a function of initial state energy (E_i) for three difference values of photon energies $\mathrm{h}\omega$ = 1.97283 eV, 2.92522 eV and 3.55109 eV respectively. Here the value of applied electric field (F) is equal to 1.6 x 10^{11} V/m. and initial state energy E_i is chosen below the Fermi level (E_F = 0.0). From the plot, we have seen that PFEC for photon energy $\hbar\omega = 1.97283$ eV shows a peak at initial state energy E_i = 2.45 eV below the Fermi level. With the further decrease in initial state energy PFEC also decreases and minimum at around E_i = 2.65 eV below the Fermi level. We have also plotted in the same Fig. 4 the calculated results of the

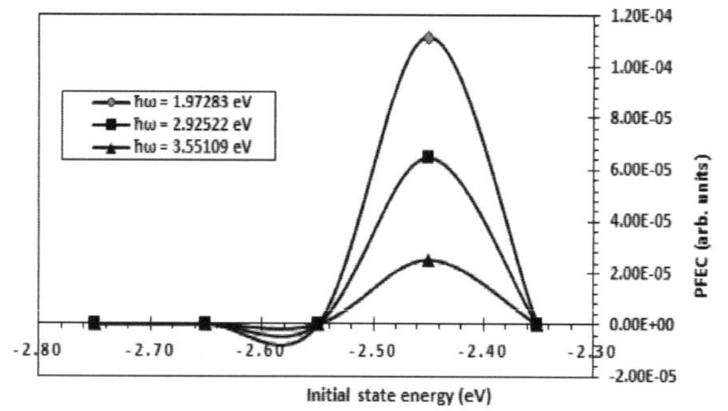

Figure 4. Plot of variation of PFEC against the initial state energy for three values of photon energies. The Fermi energy $E_F = 0.0$ is taken as reference level and the angle of incident photon radiation is $\theta_i = 45^0$.

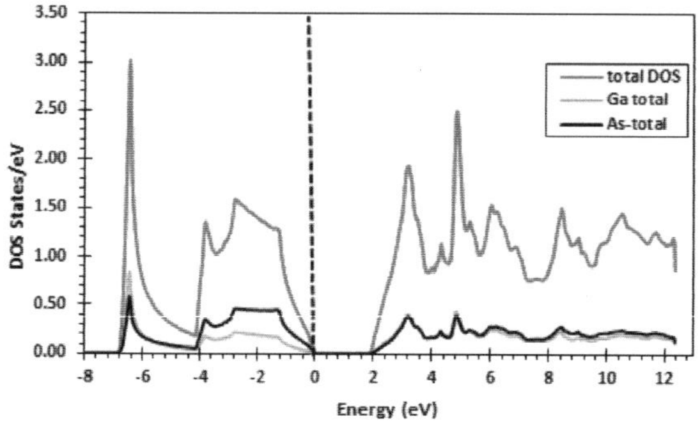

Figure 5. Plot of Total DOS for GaAs, Ga and As. Fermi level is at 0.

PFEC for photon energies $\hbar\omega = 2.92522$ eV and 3.55109 eV respectively. From this plot, we have also seen that maxima in PFEC occur at same initial state energy that is, E_i = 2.45 eV for both photon energies 2.92522 eV and 3.55109 eV below Fermi level respectively. Here the peak in PFEC is higher for low photon energy $\hbar\omega = 1.97283$ eV than for photon energies 2.92522 eV and 3.55109 eV. Therefore, the large value of PFEC for low photon energy value $\hbar\omega = 1.97283$ eV whereas low value of PFEC for high photon energy $\hbar\omega = 3.55109$ eV. This is the reason that PFEC is inversely proportional to cube of frequency of incident photon radiation as given in Eq. (1). In all the three cases we find that with the further lowering of initial state energy beyond the peaks occurrence, the values of PFEC decreases exponentially.

The total DOS plot for GaAs, Ga and As is shown in fig. 5. In this plot Fermi level is taken at 0. The maximum peak occurs at 6.45 below the Fermy level is neglected due to occurrence of peak in core region. The second maxima in total DOS of GaAs occurs at 2.77 eV in valence region below the Fermi level. In valence region most of the contribution are coming from *s* and *p*-states electrons of Ga and *p*-state electrons of As atoms. However to compare the contribution of Ga and As atoms, *p*-state electrons of As atom is more contributing than *p*-state electrons of Ga atom. This is evident from partial DOS plots of GaAs [6].

The origin of peaks occurred at 2.45 eV below the Fermi level as shown in Fig. 4 can be correlated with the occurrence of the maxima at 2.77 eV in the total DOS plot for GaAs, Ga and As which is shown in Fig. 5. This means that the origin of peaks at 2.45 eV below the Fermi level as shown in Fig. 4 is due to contribution by *s* and *p*-states of Ga and *p*-state of As. This is evident from the partial DOS plots of Ga and As atoms where the maximum in DOS was contributed by Ga-*s* and *p* and As-*p* states electrons [6].

4. Conclusions

In calculating PFEC we have used the Kronig-Penney potential model which have been used by Thapa and Kar [1]. It is found that the behaviour of PFEC as a function of applied field and initial state energy shows similar trends also in the case of GaAs. The occurrence of peak in PFEC at initial state E_i = 2.45 eV below the Fermi energy can be addressed due to band structure effects. This is evident from the plots of density of state (DOS) in GaAs which is shown in Fig. 5. However the discrepancy in the location of peaks as shown in Fig. 3 and Fig. 4 can be attributed to the choice of the wavefunctions to evaluate the matrix element. However, few drawbacks are still existing which must be attended for better accuracy. For example, we have used the real and imaginary dielectric constants which had been calculated by using the density functional theory. This was included to the model as proposed by Bagchi and Kar [3]. We are trying to correlate appropriately the dielectric model in the context of density function theory to calculate vector potentials for transition probability.

Acknowledgments
I am grateful to Mizoram University UGC fellowship, RKT acknowledges a grant from SERB (DST, India) via grant No. EMR/2015/001407, Dt.11th March 2016.

References
[1] Thapa R K and Kar N 1988, "Photoemission calculation from band States using Kronig-Penny model and spatially varying photon field," *Indian J. Pure. Appl. Phys.* **26** 620-623
[2] Gao Y and Reifenberger R 1987, "Photofield emission from transition-metal surface states," *Phys. Rev.* B **35** 4284-4290
[3] Bagchi A and Kar N 1978, "Reflection effects in angle-resolved Photo emission from surface states on metal," *Phys. Rev.* B **18** 5240-5247
[4] Blaha P, Schwarz K, Madsen G K H, Kvasnicka D and Luitz J 2015, "An augmented plane wave plus local orbitals program for calculating crystal properties, in: WIEN2K", *Vienna Universitat of Technology*, Vienna
[5] Thapa R K and Gunakar Das 2005, "A simple theory of photofield emission from the surface of a metal," *Intl. Jour. Mod. Phys.* B **19** 3141-3149
[6] Nirmala devi L, and Thapa R K, "study of electronic and optical properties of Gallium arsenide (GaAs) by using Full-Potential Linearized Augmented Plane Wave Method (FP-LAPW), "unpublished"
[7] Nirmala devi L and Thapa R K 2016, "Study of Photofield emission in GaAs using Kronig-Penney model," *Journal of Materials Science and Engineering*, A **6 (3-4)** 110-112
[8] Thapa Ram Kumar, Nirmala Devi L, Benjamin Vanlalruata, Rebecca Lalngaihawmi, Amit Shankar, Sandeep Chettri, Dibya Prakash Rai and Ghimire M P 2015 *Proc. 2015 International Conference on Electrical and Electronics: Techniques and Applications (EETA 2015)* (Phuket, Thailand) pp 53-57

Study of structural and electronic properties of fluoride perovskite $KMgF_3$ using FP-LAPW method

K Neupane [1*] and R K Thapa [1]

[1]Department of Physics, Condensed Matter Theory Research Group, Mizoram University, Aizawl 796 004-Mizoram

Corresponding Author:kneupane22@gmail.com

Abstract. In this paper, we present the electronic properties of fluoride perovskite $KMgF_3$ by using the full potential linearized augmented plane wave (FP-LAPW) method. The densities of states and energy band structures have been calculated by using the generalized gradient approximation (GGA) as exchange-correlation potential. The results obtained are in agreement with the previous experimental and theoretical results. The obtained value of energy band gap of 7.2 eV indicates the insulating behaviour of the sample material.

1. Introduction

The ternary fluoro-perovskite like $KMgF_3$ has great potential for a variety of device applications in optical, ferroelectric, antiferromagnetic systems due to their wide band gaps [1-2]. It is always an advantage to know the physical and electronic properties of such order to understand their possible applications. Perovskites are well known for their applications in different fields of science and technology because of their wide range of electro- optic, mechanical, semiconducting and insulating properties. $KMgF_3$ is a technologically important fluoro-perovskite which finds applications as a vacuum- ultraviolet-transparent material for lenses in optical lithography steppers in electro- optical applications [3-5]. It is desirable for scintillators and radiation dosimeters when it is doped approximately [6]. Neupane *et al.* [7] have reported an energy band gap of 6.8 eV for $RbCaF_3$ from the FP-LAPW method based calculation, which also suggests an insulating behaviour of this material.

In this paper, we will do the theoretical investigations of perovskite-type fluoride $KMgF_3$. For this purpose for exchange correlation, generalized gradient approximation (GGA) is employed which is implemented in wien2k code [8].

2. CRYSTAL STRUCTURE AND COMPUTATIONAL DETAILS

The unit cell of fluoro-perovskite $KMgF_3$ with space group (Pm-3m) contains three independent atoms that form the cubical structure. The atoms of $KMgF_3$ are located at the Wyckoff positions K(0.0,0.0,0.0), Mg(0.5,0.5,0.5), F1(0.0,0.5,0.5), F2(0.5,0.0,0.5) and F3(0.5,0.5,0.0) [10] to form the crystal structure as shown in figure1.

The generalized gradient approximation (GGA) as proposed by Perdew *et al.* [9] has been used to correct the exchange-correlation effect. For volume optimization of $KMgF_3$, we have used the experimental lattice constant a=3.979 Å [10] and the optimized volume versus energy curve is shown in Fig.1. This optimized lattice constant is used to study the density of state (DOS) and band energy of $KMgF_3$ by using the full potential linearized augmented plane wave (FP-LAPW) method of KS-DFT [11].We used here GGA to describe the electron exchange and correlation potential. Non spherical contributions to the charge density and potential within the muffin tin (MT) spheres are considered and the cut-off parameter is $R_{MT} \times K_{max} = 7$ where K_{max} is the maximum value of the reciprocal lattice vector in the plane wave expansion and RMT is the smallest atomic sphere radii of all atomic spheres. In the interstitial region, the charge density and potential are expanded as a Fourier series with wave vectors up to G_{max}=12 a.u^{-1}. The number of k-points used in the irreducible part of the Brillouin zone is 1000. The criterion for the convergence of the self-consistent DFT calculation is 0.0001 Ry in total energy. However the core states are treated relativistically, the semi-core states are treated semi-relativistically by ignoring the spin-orbit (SO) coupling.

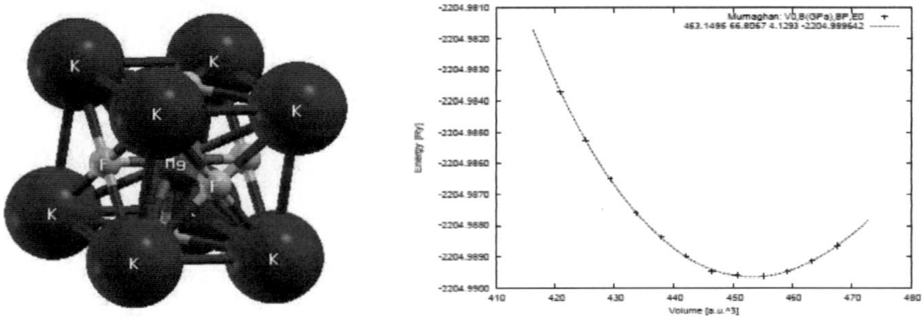

Figure 1. (Left) Crystal structure of $KMgF_3$, (right) volume optimization curve.

3. RESULT AND DISCUSSIONS

3.1. *Crystal structure*

The variation in the total energy as a function of volume is shown in figure1.The calculated value of the equilibrium lattice constant (a) is found as 4.0646 Å. The energy versus volume data was fitted to a Murnaghan equation of state [12] to obtain the bulk modulus (B) and its first pressure derivative (B'). Our calculated values of lattice constant, B, and B' are compared with previous experimental and theoretical results which are shown in table 1. We note that the result of the lattice constant obtained is slightly higher than the experimental values [10, 15].

Table 1.

	Lattice Constant a (in Å)	Bulk Modulus B (in GPa)	Pressure Derivative B'
This study	4.0646	66.8067	4.1293
Other, GGA	4.081 [13]	72.01 [13]	4.65[13]
	4.040 [14]	72.15 [14]	4.68 [14]
Experimental	3.989 [15]	70.40 [15]	
	3.979 [10]		

3.2 *Density of states (DOS)*

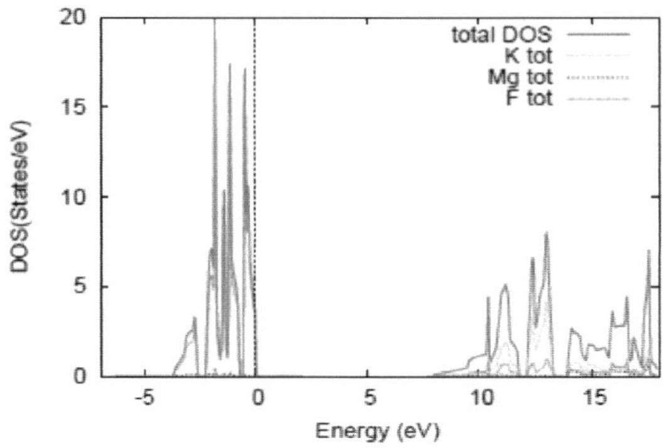

Figure 2. Plot of total DOS for $KMgF_3$, K, Mg and F (Fermi energy is at zero)

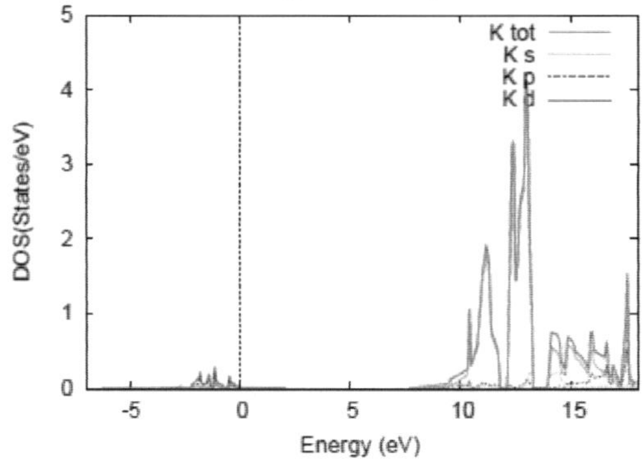

Figure 3. Plot of total and partial DOS for K Fermi energy is at zero)

Figure 2 shows the plot of total density of states (DOS) of $KMgF_3$ and individual atoms K, Mg and F respectively. The maximum peak and other two narrow peaks are occurred at 1.82 eV, 1.14 eV and 0.45 eV respectively in valence region below the Fermi level. In the case of total DOS, contribution of the individual atoms in the valence region, we find only the contribution due to F atom which is accountable in the valence region. The total contribution by K and Mg atoms are more or less negligible in the valence region which is evident from the small hump in DOS. In the conduction region above the Fermi level in figure 2, we find the value of DOS is maximum at 13.07 eV due to the main contribution of K atom and small peaks in DOS occur from 10.45 eV to17.50 eV due to the contribution of K, Mg and F atoms.

From the total and partial DOS plots of K atom as shown in figure 3, we find no sharp peaks in valence band but only *p* and *d* state electrons are contributing to the DOS in negligible height. However, in the conduction band, the highest peak in DOS is contributed by *d* state electron of K atom at 13.0 eV and other smaller peaks are observed from 10.33 eV to 17.55 eV due to the main contribution of *d* state electrons of K atom.

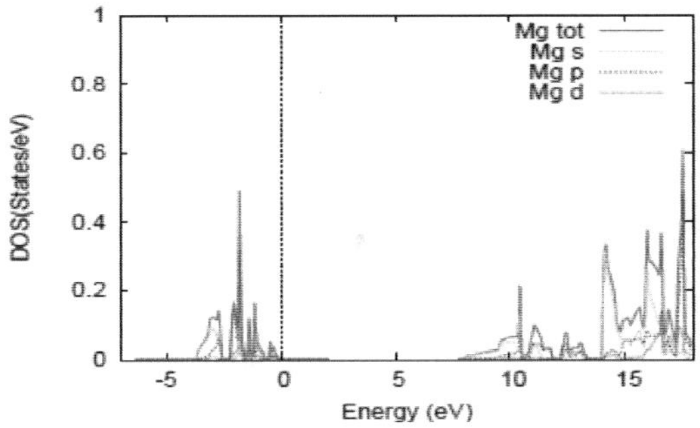

Figure 4. Plot of total and partial DOS for Mg (Fermi energy is at zero)

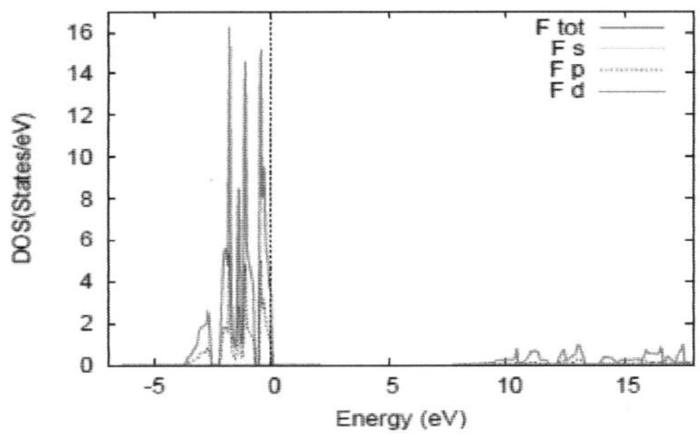

Figure 5. Plot of total and partial DOS for Mg (Fermi energy is at zero)

From the total and partial DOS of plots of Mg atom as shown in figure 4, we find a narrow peak at 1.85 eV in the valence band due to the *p* and *d* state electrons and small peaks are observed from 0.43 eV to 2.93 eV due to the contribution of *s*, *p* and *d* state electrons below the Fermi level. However, in the conduction band, *p* state electron of Mg atom contributes to the occurrence of peak at 17.55 eV and small peaks are observed from 10.33eV to 17.55eV due to the contribution of *s*, *p* and *d* state electrons of Mg atom.

Figure 5 shows the total and partial DOS of plots of F. In the valence band, there is a sharp peak occurring at 1.82 eV and other two small peaks are observed at 1.14 eV and 0.45 eV respectively below the Fermi level due to the *p* state electrons and virtually with no contribution by *s* and *d* state electrons. However, in the conduction band, there are very small peaks occur due to the *s*, *p* and *d* state electrons of F atom.

3.2. Band structures

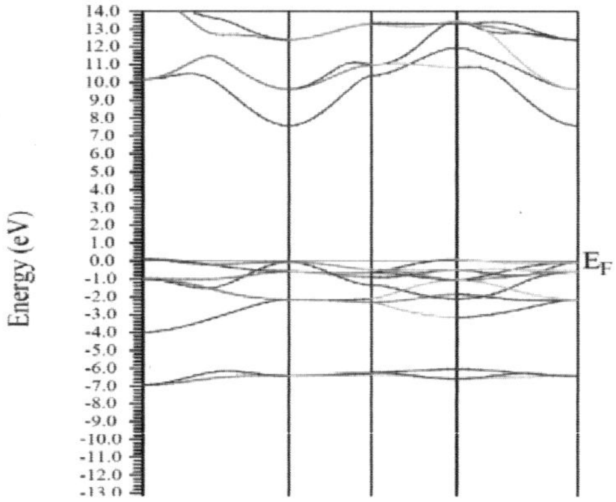

Figure 6. Electronic band structure of $KMgF_3$ along the high-symmetry directions of the first Brillouin zone.

The calculated electronic band structure for fluoro-perovskite $KMgF_3$ along the high-symmetry directions of the Brillouin zone is shown in figure 6. In the valence band (figure 6), the lowest lying band has been found to occur at 9.5 eV below Fermi level due to the core state electrons of K, Mg and F atoms. We also observe from figure 6 that the maximum band energy occurs at the Fermi level at the symmetry point R. In the conduction band, minimum in energy occurs at 7.2 eV above the Fermi level at the point symmetry and from this plot in figure 6, we find that it is an indirect type of transition which takes place along R - symmetry directions. The calculated value of indirect band gap is 7.2 eV, which is much smaller than the experimental value 12.4 eV [16] for $KMgF_3$.

4. CONCLUSIONS

From the volume optimization in figure1, we have found the optimized lattice constant, pressure, pressure derivative and total energy are as 4.0646 Å, 66.8067 GPa, 4.1293 and -2204. 9896 Ry respectively. From the total DOS plots of $KMgF_3$ as given in figure 2, we find that maxima in peaks in the valence region are due to only F atom. This is also evident from the partial DOS plots of K, Mg and F atoms as given in figure 3, figure 4 and figure 5 respectively. We find the maxima in peak is due to *p*-state electrons of Mg and F atoms in the valence region of partial DOS of Mg and F atoms. The peak heights in partial DOS due to electrons from K atom in the valence region is negligible. Similarly in the conduction region, it is only *d*-state electrons of K (figure 3) which is a main contribution to the maxima in the total DOS of $KMgF_3$.We have found from our study (figure 6) that the band gap of $KMgF_3$ is 7.2 eV which is large and hence $KMgF_3$ is an insulator.

References

[1] Ouenzerfi R E, Ono S, Quema A, Goto M, sakai M 2004 *J. Appl. Phys.***96** 7655
[2] Zhang F, Mao Y, part T J, Wong S S 2008 *Adv. Funct. Mater.* **18** 103
[3] Nishimatsu T N, Terakubo N, Mizuseki H, Kawazoe Y, Pawlak D A,Shimamura K, Ichinose N, Fukuda T 2003 *Jpn. J. Appl. Phys.* **42** 5082
[4] Ho¨rsch G, Paus H J 1986 *Opt. Commun.***60** 69
[5] Fukuda T, Shimamura K, Yoshikawa A, Villora E G 2001 *Opt. Elect. Rev.* **9** 109
[6] Furetta C, Santopietro F, Sanipoli C, Kitis G 2001 *Appl Radiat Isot* **55** 533
[7] Neupane K Thapa R K 2016 *IJEART* 2454-9290 **2** 31-33
[8] Blaha P, Schwarz K, Madsen G K H, Kvasnicka D, Luitz J, Schwarz K 2012 *An Augmented Plane Wave plus Local Orbitals Program for Calculating Crystal Properties* Wien2K User's Guide Techn. Universitat Wien,Wien
[9] Perdew J P, Burke F S I, Borodi G 1996 *J. Cryst. Growth.* **169** 89
[10] Darabont A, Neamtu C, Ouenzerfi R E, Ono S, Quema A, Goto M, sakai M 2004 *J. Appl. Phys.* **96** 7655.

[11] Kohn W, Sham L J 1965 *Physical Review* **A140** 1133
[12] Murnaghan F D 1944 *Proc Natl Acad Sci* **30** 244
[13] Vaitheeswaran G, Kanchana V, Kumar R S, Cornelius A L, Nicol M F, Svane A, Delin A, Johansson B 2007 *Phys. Rev.* **B 76** 014107
[14] Pilania G, Sharma V 2013 *J. Mater. Sci.* **48** 7635-7641
[15] Rosenberg H M, Wigmore J K 1967 *Phys. Lett.* **A 24** 317
[16] Heaton R A, Lin C C 1982 *Phys. Rev.* **B 25** 3538

Analysis of photoassisted field emission current in Iridium and Titanium by using Kronig-Penney potential model and transfer Hamiltonian method

Rosangliana[1]* and R. K. Thapa[2]

[1]Department of Physics, Govt. Zirtiri Residential Science College, Aizawl 796001, Mizoram, India
[2]Department of Physics, Mizoram University -796004, India

*Email: rslct7@gmail.com

Abstract. We present here a model calculation of photoassisted field emission current (PFEC) by using transfer Hamiltonian method. Matrix element for the transition probability is calculated by using the wavefunctions which are deduced by applying the Kronig-Penney potential model PFEC is calculated for Iridium and Titanium, and the results are compared with experimental data.

Keywords : 73.20.-r, 73.30.+y, 79.70.+q

1. Introduction

Photoassisted field emission is a technique in which a metal is irradiated by an incident laser radiation of photon energy ($\hbar\omega$). Photon energy is usually less than the work function (ϕ) of the metal. The incident radiation photoexcites the electrons to states which lies between the Fermi level and the vacuum level, hence these electrons are confined within the metal surface. A strong static electric field (~ 10^{11} Vm^{-1}) when applied to the surface of the metal causes the photoexcited electrons to tunnel through the surface potential barrier into the surrounding vacuum. These electrons which are now emitted into the vacuum region constitute the measurable current called photoassisted field emission current (PFEC). One attractive application of photoassiated field emission (PFE) is to allow short time modulation of laser to combine photoemission and field emission which can control the duration of emission and increase the emitted peak current. This had been possible in the case of ZrC single tip and nanoscale current pulses have been emitted with laser pulses at 1064 nm illuminating a ZrC tip under high DC electric field [1]. We present here the calculation of PFEC from Iridium and Titanium by using the transfer Hamiltonian method. Matrix element for transition $M_{fi} = \langle f|\mathbf{A}.\mathbf{p}+\mathbf{p}.\mathbf{A}|i\rangle$ had been evaluated by using the Kronig-Penney periodic potential [2] for photoemission studies. The calculated result of PFEC from Iridium and Titanium are compared with the experimental data as measured by [3] and [4].

2. Theory

The transfer Hamiltonian (TH) is based on the division of the tunnelling system into two subsystems. The transfer of electrons between the two subsystems is described by a term called the transfer Hamiltonian, from which the method derived its name. The separation of the system into two subsystems is only possible if the coupling between the two subsystems is weak and the probability of an electron tunnelling through the potential barrier is small [5]. The transfer Hamiltonian method is the approximate separation of the Hamiltonian for the tunnelling system into its components which is treated as a perturbation, leading to a total Hamiltonian. The total Hamiltonian of electrons in one dimension is

$$H = T + V_m\theta(-z) + V(z) - eFz\theta(z) \qquad (1)$$

Here, T is one electron kinetic energy operator, V_m the periodic metal potential, $V(z)$ the outside potential in the absence of field and F is the high static electric field. Here $V(z) = V_o - eFz - (e^2/4z)$ where V_o is the height of the potential barrier and $-\frac{e^2}{4z}$ is the image potential.

A p-polarized radiation of photon energy $\hbar\omega$ is incident on an electron lying at energy E_i below the Fermi level as shown in figure 1. This incident radiation, usually a laser beam causes the transition of electron from the initial state ψ_i to the final state ψ_f. As this electron still lies below the vacuum level, a high electric field (~ 10^{11} $\mathrm{Vm^{-1}}$) will now cause the electron to be transmitted through the potential barrier to vacuum region thereby causing photoassisted field emission.

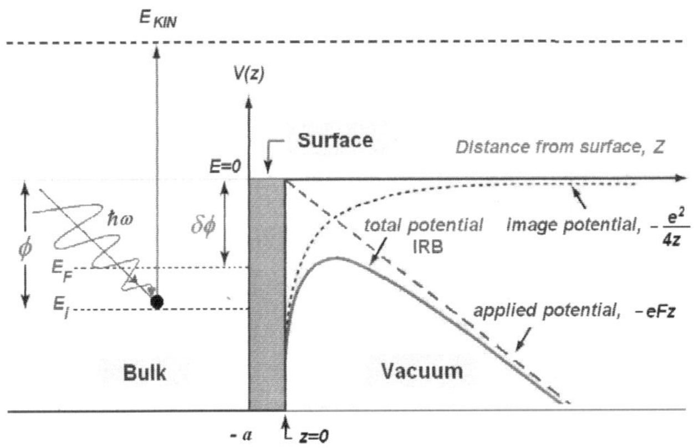

Figure 1. Model potential used for PFE calculations for describing the band states with a surface of width *a*.

This happens as the high static field causes the reduction of the work function of the metal causing *Schottky effect* which brings in the image rounded barrier (IRB). This will reduce the height of the step potential and the work function at the surface as shown in figure 1.

Considering the transfer Hamiltonian method, the transition matrix between the two states ψ_i and ψ_f is given by

$$M_{fi} = \frac{\hbar^2}{2m}\int_{-\infty}^{+\infty}\left[\psi_f^*\frac{\partial\psi_i}{\partial z} - \psi_i\frac{\partial\psi_f^*}{\partial z}\right]\mathrm{d}s - \frac{\hbar^2 e^2}{8m}\int_{-\infty}^{+\infty}\left[\psi_f^*\left(\frac{1}{z}\right)\psi_i\right]\mathrm{d}z$$
$$+\frac{\hbar^2}{2m}\int_{-\infty}^{+\infty}\left[\psi_f^*\left(A_\omega(z)\frac{\mathrm{d}}{\mathrm{d}z} + \frac{1}{2}\frac{\mathrm{d}A_\omega(z)}{\mathrm{d}z}\right)\psi_i\right]\mathrm{d}z \qquad (2)$$

One dimensional Kronig-Penney potential is used [2] to deduce ψ_i in equation (2). In one dimension, this is given by

$$\psi_i(z)=\begin{cases}\left(\dfrac{m}{2\pi\hbar^2 k_i}\right)^{1/2}\begin{bmatrix}\left(1-\mathrm{i}P\exp(-\mathrm{i}\delta)\mathrm{i}\sin\delta\right)\exp(\mathrm{i}k_i z)\\-\left(P-\mathrm{i}\exp(\mathrm{i}\delta)\sin\delta\right)\exp(-\mathrm{i}k_i z)\end{bmatrix}, & for\ z\le 0\\ \left(\dfrac{m}{2\pi\hbar^2 k_i}\right)^{1/2} T\exp(-\chi z), & for\ z>0\end{cases}\tag{3}$$

The reflection and the transmission coefficients across the boundary plane P and T are deduced by using boundary condition and are

$$P=-\frac{\left[(\chi-\mathrm{i}k_i)-(k_i-\chi)\exp(\mathrm{i}\delta)\sin\delta\right]}{\left[(\chi-\mathrm{i}k_i)+(k_i-\chi)\exp(\mathrm{i}\delta)\sin\delta\right]}$$

and
$$T=\frac{\left[2k_i\sin 2\delta\right]}{\left[(\chi-\mathrm{i}k_i)+(k_i-\chi)\sin\delta\exp(\mathrm{i}\delta)\right]}\tag{4}$$

where $\delta=-\tan\left(\dfrac{\hbar^2 k_i}{mg}\right)$, is the phase shift introduced in the transmitted wave and g is the strength of the potential. The one dimensional final state wave function used in equation (2) is

$$\psi_f(z)=\begin{cases}\left(\dfrac{m}{2\pi\hbar^2 q}\right)^{1/2}\left[\dfrac{2q}{(q+k_f)}\right]\exp\left(\mathrm{i}k_f z\right)\exp\left(-\alpha|z|\right), & for\ z\le 0\ \ (Bulk\ and\ surface)\\ \left(\dfrac{m}{2\pi\hbar^2 q}\right)^{1/2}\left[\exp\left(\mathrm{i}qz\right)+\left\{\dfrac{(q-k_f)}{(q+k_f)}\right\}\exp\left(-\mathrm{i}qz\right)\right], & for\ z>0\ \ (Vacuum)\end{cases}\tag{5}$$

$$Here,\quad k_i=\left(\frac{2mE_i}{\hbar^2}\right)^{1/2}\qquad k_f=\left(\frac{2mE_f}{\hbar^2}\right)^{1/2}$$

$$q=\left[\left(\frac{2m}{\hbar^2}\right)(E_f-V_o)\right]^{1/2}\qquad \chi=\left[\left(\frac{2m}{\hbar^2}\right)(V_o-E_i)\right]^{1/2}\tag{6}$$

$$where,\ E_f=E_i+\hbar\omega\ \ and,\ V_o=E_f+\phi.$$

The term $e^{-\alpha|z|}$ is responsible for the inelastic scattering of the electrons on the surface and bulk side. The one dimensional form of vector potential of interest used in equation (2) for the bulk $(z<-a)$, surface $(-a\le z\le 0)$ and vacuum $(z>0)$ regions are given by [8]

$$A_\omega(z)=\begin{cases}A_1, & for\ z\le -a\quad (Bulk)\\ \dfrac{A_1\varepsilon(\omega)a}{\left(a+[1-\varepsilon(\omega)]z\right)}, & for\ -a\le z\le 0\ \ (Surface)\\ A_1.\varepsilon(\omega), & for\ z\ge 0\quad (Vacuum)\end{cases}\tag{7}$$

where, $A_1=-\dfrac{\sin 2\theta_i}{\left[\left(\varepsilon(\omega)-\sin^2\theta_i\right)^{1/2}+\varepsilon(\omega)\cos\theta_i\right]}$ is a constant depending on the dielectric function $\varepsilon(\omega)$ and angle of incidence θ_i.

Photoassisted field emission current density can be calculated by using the formula [6]

$$\frac{\mathrm{d}j(E)}{\mathrm{d}E}=\left(\frac{2\pi e}{\hbar}\right)|M_{fi}|^2 D(W)\delta\left(E_f-E_i\right)\delta\left(E-E_i\right)f\left(E_i\right)\tag{8}$$

The transmission probability $D(W)$ used in equation (8) had been calculated by [7] which is

$$D(W)=\left[\frac{\left(W^{1/4}\pi^{1/2}\right)}{\left(\hbar eF\right)^{1/6}}\right]\left[\frac{2ik_i}{\left(ik_i+\chi\right)}\right]\left(2m\right)^{1/12}\times\exp\left[-i\left\{\left(\frac{2W^{3/2}\left(2m\right)^{1/2}}{3\hbar eF}\right)+\frac{\pi}{4}\right\}\right]. \quad (9)$$

As the expansion in equation (8) cannot be solved analytically, PFEC is evaluated by writing FORTRAN programs.

3. Results and discussions

Formalism developed for calculating the photoassisted field emission current (PFEC) by using the transfer Hamiltonian method is applied to the case of Ir and Ti. We have calculated PFEC as a function of applied field, location of electron states with reference to a Fermi level for surface width a = 10 a.u. and narrow surface width $(a \approx 0)$. As it is a theoretical calculation, we have assumed the following data both for Ir and Ti.

Potential barrier height (V_0)	= 15 eV,
Work function (ϕ)	= 4.928 eV,
Fermi energy (E_F)	= 10.06 eV,
Scattering factor (α)	= 0.35
Angle of incidence	= 45°

The experimental values of dielectric constants $\varepsilon(\omega)$ had been used for evaluating the vector potential $A_\omega(z)$ given by equation (7).

3.1 Iridium

Plot of PFEC as a function of applied electric field for a = 10 a.u., is shown in figure 2. Here we have shown the plots of PFEC for the photon energies $\hbar\omega$ =1.96 eV and 2.70 eV. It is found that as the values of applied field is increased, PFEC decreases from a high value in an exponential manner for both the values of photon energies 1.96 eV and 2.70 eV respectively. The exponential decrease in the PFEC is due to the presence of exponential term in transmission probability [3] have also experimentally measured PFEC in the case of Iridium and compared with their calculated true photocurrent results. It was found that PFEC decreases from a high value towards zero when the applied field is increased both for photon energies 1.96 eV and 2.70 eV respectively. The measured PFEC decreases but not exponentially in both the cases [3].

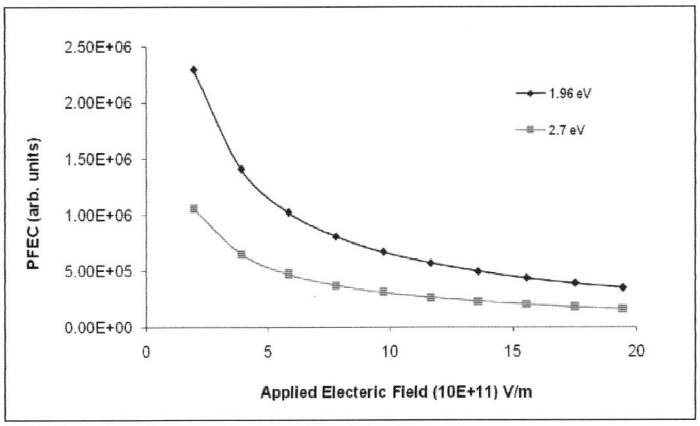

Figure 2. Plot of PFEC against applied field F in Iridium for photon energies 1.96 eV and 2.70 eV. Initial state energy E_i =1eV below Fermi level, and angle of incidence of photon radiation is 45°, surface width a = 10 a.u.

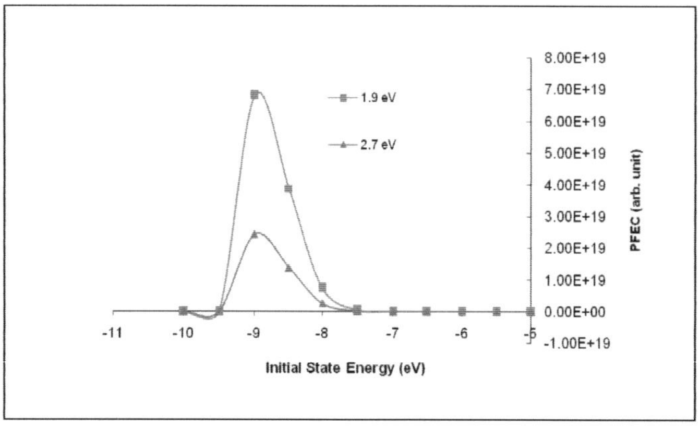

Figure 3. Plot of PFEC against initial state energy E_i below Fermi level (E_F =0.0) in Iridium for photon energies 1.96 eV and 2.70 eV. Here value of applied field $F = 3.08\times10^{11}$ Vm^{-1} and angle of incidence of photon radiation is 45°, surface width $a =$ 10 a.u.

Plot of PFEC as a function of initial state energy of the electron with respect to the Fermi level is shown in figure 3. PFEC is plotted for a = 10 a.u. for photon energies 1.96 eV and 2.70 eV respectively. For both the photon energies, peak in PFEC occurs at initial state energy -9 eV below E_F (Fermi level). It is seen from the figure that peak is higher for 1.96 eV photon energy than 2.70 eV. Also occurrence of peaks for both photon energies at the same location of initial state energy i.e., -9 eV below E_F. This is attributed to the contribution from the *d*-electron states in the valence band. In figure 4, we have shown the plots of matrix element $M_{fi} = \langle f | A.p + p.A | i \rangle$ and calculated PFEC as a function of incident photon energies. Both M_{fi} and PFEC increases as photon energy of the incident laser radiation increases and then showed a maxima at 2 eV photon energy. Beyond 2 eV, both showed decrease in the magnitude for M_{fi} and PFEC.

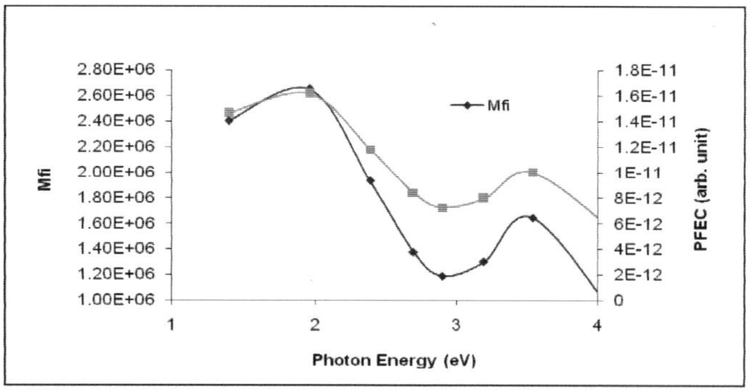

Figure 4. Plot of matrix element M_{fi} and PFEC against photon energy in Iridium for values of applied field $F = 3.08\times10^{11}$ Vm^{-1}, initial state energy $E_i = 1\,eV$ Fermi level. The angle of incident photon radiation is 45° and surface width a = 10 a.u.

At 3.5 eV, a second peak is observed this is due to the effect of applied field in the surface. Similar trends in the behaviour of M_{fi} and PFEC is expected because M_{fi} is involved directly in the formula for calculating PFEC in equation (8).

3.2 Titanium

Plot of variation of PFEC as a function of applied field for surface width a = 10 a.u. is shown in figure 5. Plot is shown in the case of photon energies 1.96 eV and 2.70 eV respectively. Here also we find that just like in the case of Iridium, the PFEC decreases from a high value as applied field increases. The decrease in PFEC is exponential in nature as in the case of Iridium. Reference [4] has also shown experimentally that PFEC decreases from high value toward minimum as applied field is increased. Calculated results as obtained by [4] also showed similar trends.

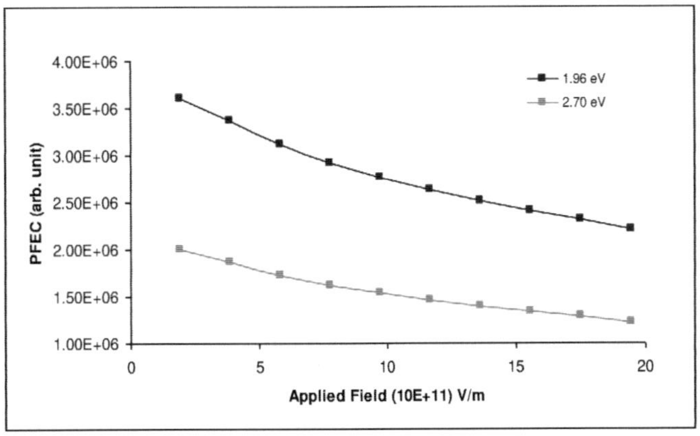

Figure 5. Plot of PFEC against applied field F in Titanium for photon energies 1.96 eV and 2.70 eV. Initial state energy E_i = 1.0 eV below Fermi level, and angle of incidence of photon radiation is 45°, surface width a = 10 a.u.

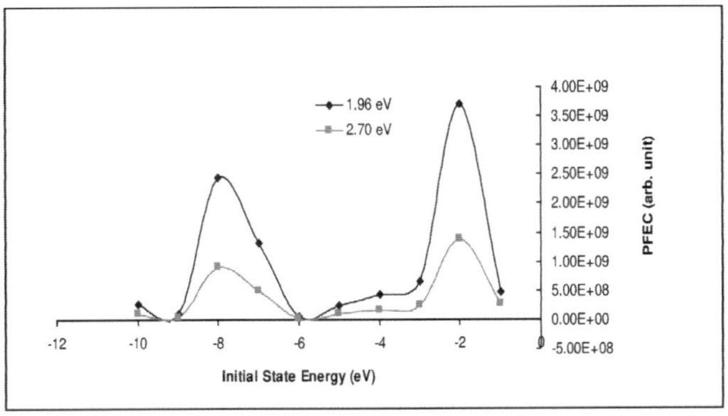

Figure 6. Plot of PFEC against initial state energy E_i below Fermi level (E_F =0.0) in Titanium for photon energies 1.96 eV and 2.70 eV. Here value of applied field $F = 3.08 \times 10^{11}$ Vm^{-1} and angle of incidence of photon radiation is 45°, surface width a = 10 a.u.

In figure 6, plot of PFEC against initial state energies of electron are shown for two photon energies 1.96 eV and 2.70 eV respectively. Similar features in the variation of PFEC is observed with the occurrence of single peak in PFEC at initial state energy -2 eV below E_F both for photon energies 1.96 eV and 2.70 eV. For the same photon energies, peaks in PFEC is also obtained at -8 eV below E_F, but of lower heights than at -2 eV energy below E_F.

The surface density of states for Ti as measured experimentally [4] for p-polarized photon energy 1.96 eV showed several peaks below E_F with a maximum at -0.15 eV below E_F. Occurrence of peaks in PFEC at -2 eV agrees qualitatively with the experimental results of [4]. In figure 7, plots of M_{fi} and PFEC as a function of photon energy is shown. Just like in the case of Ir, variation of M_{fi} and PFEC showed similar behaviour as photon energy increases. Peaks in M_{fi} and PFEC occurs at 0.8 eV, after which plots in M_{fi} and PFEC decreases with the further increase of photon energies. The similar behaviour found in the plot of M_{fi} and PFEC as a function of photon energy in both the metals is due to the inclusion of vector potential A in the matrix element.

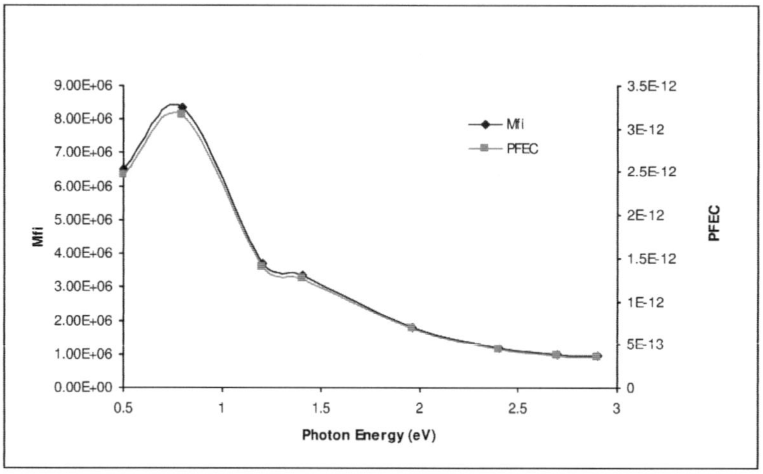

Figure 7. Plot of matrix element M_{fi} and PFEC against photon energy in Titanium for values of applied field $F = 3.08\times10^{11}\,\text{Vm}^{-1}$, initial state energy $E_i = 1\,\text{eV}$ Fermi level. The angle of incident photon radiation is 45° and surface width a =10 a.u.

We have also checked the effect of inclusion of surface width on PFEC. For this purpose we have included a narrow surface width $(a \approx 0)$ and calculated PFEC as a function of applied field. It is seen that PFEC increases drastically to a high value without showing exponential type of variation as in the case of results obtained with surface width a = 10 a.u.

4. Conclusions

We have shown the variation of PFEC as a function of applied electric field and photon energy by considering the transfer Hamiltonian method. Exponential decrease in PFEC both in the case of Ir and Ti shows features which were also found in the case of tungsten [9]. Calculated results of PFEC also showed similar behaviour [3] in the case of Ir and [4] in the case of Ti. Here we have calculated band structures of Ti by using the FP-LAPW method [6] which is implemented in the WIEN2k code [10].

In the case of Ir, occurrence of peaks at -9 eV below Fermi level is shown in figure 3. This is attributed to the surface related features. Due to the high value of work function of Iridium; it was difficult to obtain accurate data at the lower photon energies.

Further increase in applied field does not bring any change in PFEC but instead keeps it minimum. As evidence of our calculations, we find that matrix element when plotted as a function of photon energies is not constant [11]. We have also not found oscillations in PFEC as shown by [12] in their experimental results.

There are drawbacks in the method of calculations which must be taken into considerations. For example, the wavefunctions ψ_i chosen are too simple. For this, we must derive ψ_i with correct type of potential for the bulk and the surface regions. The wavefunctions appropriately chosen should have the symmetry well defined. Same wavefunction is used for bulk and surface regions. Further, vector potential used should be correctly defined for both the bulk and surface regions. However, even with all these shortcomings, results obtained are comparable to other results [13].

Acknowledgment
RKT acknowledges a research grant from DAE (BRNS), RSL thanks Department of Physics, Mizoram University for permission to use the computational facilities.

References
[1] Ganter R, Bakker R J, Gouch C, Paraliev M, Padrozzi, Le Pimper F, Rivkin L, Wrulish A. 1995 *Nucl, Inst. Methods in Physics Research,* A**565**, 423-29
[2] Thapa R K, and Kar N, 1995 *Surf. Sci.* **338**, 138-142
[3] Gao Y and Reifenberger R, 1987 *Phys. Rev.* B**35**, 4284-90

[4] Hadzel P, Radon T, Lee M J C and Chow J C L, 1999 *Surf. Sci.* **442**, 36-46
[5] Payne M C, 1986 *J. Phys. C: Solid State Phys.* **19**, 781-88
[6] Rosangliana, 2011 Theoretical study of surface effect in photoassisted field emission, Ph.D. Thesis, Mizoram University, Aizawl, India, unpublished
[7] Thapa R K and Das G, 2005 *Intl. Jour. Mod. Phys.* B**19**, 3141-49
[8] Bagchi A and Kar N, 1978 *Phys. Rev.*, B**18**: 5240-47
[9] Rosangliana, Ghimire M P, Lalmuanpuia, Sandeep and Thapa R K, 2010 *Indian Jour. Phys.* **84**, 723-27
[10] Blaha P, Schwartz K, Madsen G K H, Kvasnicka D and Luitz J, 2008 *Revised Edition, Vienna University of Technology, Vienna, Austria*
[11] Schwartz C and Schaich W L, 1981 *Phys. Rev.* B**24**, 1583-86
[12] Modinos A and Kleint Ch, 1984 *Solid State Commun.* **50**, 651-54
[13] Rosangliana, Mohanty A K and Thapa R K, 2012 *Journal of Material Science and Engineering,* B**2(2)**, 203-07

Investigation of half-metallicity of GeKMg and SnKMg by Using mBJ potential method

T. Malsawmtluanga[1], Benjamin Vanlalruata[1] R. K. Thapa[1]
Department of Physics, Condensed Matter Theory Research Group, Mizoram University, Tanhril, Aizawl, Mizoram-796004, India

tochhawng1981@gmail.com

Abstract. The electronic structures, magnetic properties and half-metallicity of GeKMg and SnKMg half- Heusler compounds have been investigated by the first- principles calculations based on the density functional theory. The spin-polarized calculations using full-potential linearized augmented plane-wave (FP-LAPW) method was utilized for the study of the compounds. The modified Becke-Johnson (mBJ) exchange potential was employed for a better description of the half metallic response of the two compounds. We have found that GeKMg and SnKMg alloys are half-metallic ferromagnets with the magnetic moment of 1 μ_B per formula unit at equilibrium lattice constant.

1. Introduction

Half-metallic ferromagnets (HMF) which shows metallic behaviour in one spin channel while the other spin channel shows semiconductor or insulator behaviour at the Fermi level have gained a lot of interest due to their due to their unique properties and potential applications in spintronic devices. Materials like these are important to scientific researchers due to their potential device applications such as non-volatile magnetic random access memories (MRAM), magnetic sensors [4-6] and of spin-injector materials because their electronic structure is metallic in only one of the two spin channels, which results in complete (100%) spin polarization at the Fermi level. The concept of half-metallic ferromagnets was first introduced by de Groot et al. [1, 2], on the basis of band structure calculations in NiMnSb and PtMnSb half-Heusler compound and since then more HM ferromagnets have been predicted theoretically or confirmed experimentally in some classes of compounds such as the half-Heusler alloys, the full-Heusler alloys, the perovskites [9], the magnetite[10-12] and the zinc-blende (ZB)[8]and wurtzite transition-metal pnictides [3]and chalcogenides [8] .

Because of their high Curie temperatures and structural similarity to conventional semiconductors with zinc-blende structure HM ferromagnets with half-Heusler and full-Heusler structures have gained research interest e.g., the full-Heusler compounds Co_2MX (M = Mn, Fe and Cr; X = Si, Ge and Sn) with $L2_1$ structure have been found to be HM ferromagnets by some theoretical and experimental research groups [35-39]. The half-Heusler compounds, CoMnX (X = P, As and Sb) [6,42] FeCrSb, FeMnSb [42] and NiCrZ (Z = P, Se and Te) and NiVAs [43-45] have also been predicted to be HM ferromagnets. Theoretically, the above-room-temperature ferromagnetism in NiCrZ and NiVAs were predicted by Şaşıoğlu et al. [45].

In this paper, we have adopted a method based on density functional theory (DFT) to investigate electronic and magnetic properties of GeKMg and SnKMg half-Heusler alloys with Cl_b-type structure by using Perdew Burke-Eenzehof generalized gradient approximation (GGA) [13]. Even though the ground state is well described by GGA, the values of the band gaps are sometimes underestimated. And so electronic properties are analysed further using the modified Becke Johnson potentiall (mBJ) calculations for improved values of the band gaps [14]. The exchange potential, introduced by Becke and Johnson, [19] was modified by Tran and Blaha.[15] The modified Becke-Johnson (mBJ) potential is given by

$$V_{mBJ}(r) = cV_x^{BR}(r) + (3c-2)\frac{1}{\pi}\sqrt{\frac{5}{12}}[2t(r)/\rho(r)]^{1/2}$$

where $V_x^{BR}(r)$ is the Becke-Roussel exchange potential, [47] *(r)* is the electron density and t(r) is the Kohn-Sham kinetic energy density. In the above equation, $c = A + B\sqrt{g}$ where g is the average of $|\nabla\rho/\rho|$ over the volume of the unit cell, and A and B are parameters adjusted to produce the best fit to the experimental values of the semiconductor band gaps. Radi A. Jishi [40] has shown that using the modified Becke-Johnson exchange potential with A = 0.4 and B = 1.0 bohr$^{1/2}$, the calculated band gaps of $MAPbI_3$, $MAPbBr_3$, $RbPbI_3$,and $CsPbX_3$ (X = Cl, Br, I) are in excellent agreement with experimental values. [41]

2. Computational method

In this paper, in order to investigate the structural stability and potential half-metallicity of half-Heusler compounds GeKMg and SnKMg we use the first-principles calculations to systematically investigate their electronic, magnetic and structural stability.

The spin-polarized density functional theory (DFT) calculations are performed with WIEN2k code [15-16]. The generalized gradient approximation (GGA) in Perdew Burke-Eenzehof generalized gradient approximation (GGA) [13,17] is used to describe the exchange correlation energy. The mBJ (modified Becke Johnson) potential [19-21] is used to get a more accurate representation of the DOS and band structures and comparisons were made with the results from PBE-GGA and mBJ calculations. A mesh of 12 × 12 × 12, consisting of 286 special k points are set in the Brillouin zone (10000 k points in the full BZ) of irreducible wedge for the integrations within the modified tetrahedron method [23] . The cut off parameter foe the plane wave was set to Kmax × Rmt= 7.0 where Kmax is the uttermost value of the reciprocal lattice vector in plane wave function and Rmt is the smallest muffin tin sphere radii. The value of Gmax is 12 where Gmax is the largest vector value in charge density Fourier expansion. These parameters ensure good convergences for the total energy.

3. Results and discussions

3.1. *Crystal Structure.*

Half-Heusler compounds, which have the chemical form ula XYZ, crystallize in the face-centered cubic C1b structure with the space group F-43m [24]. In this structure, X, Y and Z atoms occupy (1/4,1/4,1/4), (0,0,0) and (1/2,1/2,1/2) sites, respectively, and (3/4,3/4,3/4) site is empty. To get the equilibrium lattice constants of the half-Heusler alloys GeKMg and SnKMg, we calculate the total energy as a function of lattice constants for the the three possible atomic arrangements as in Table 1. The α-phase is use for calculations of the electronic and magnetic properties as it has the lowest minimized energy. Fig.2 shows optimized volume of GeKMg at equilibrium lattice constant in the α-phase and the lattice parameters at constant equilibrium is 7.26 °A for GeKMg and 7.61°A for SnKMg by using Murnaghan equation [18]. The equilibrium lattice constants, bulk modulus, energy we calculated are displayed in Table 2.

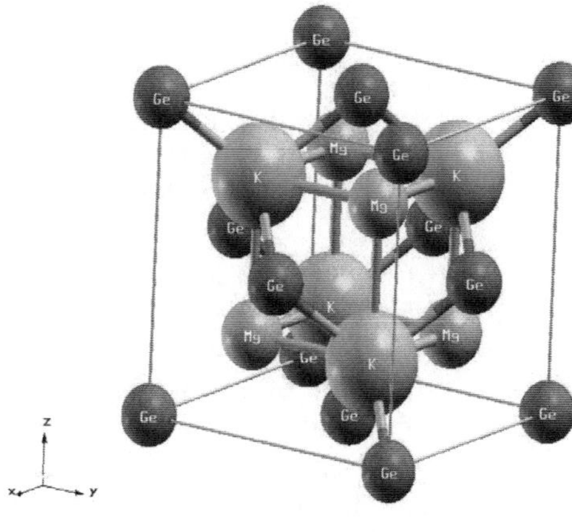

Figure 1. Unit Cell of cubic C1b-type structure for GeKMg in α-phase.

Table 1. Atomic arrangements of X,Y and Z atoms in three phases.

Phase	X	Y	Z
α	¼, ¼, ¼	0,0,0	½, ½, ½
β	0,0,0	½, ½, ½	¼, ¼, ¼
γ	½, ½, ½	¼, ¼, ¼	0,0,0

Table 2. Computed equilibrium lattice constant with bulk modulus and energy with previous calculation*[26].

Compound	GGA (A^0)	GGA (A^0)*	Bulk Modulus (Gpa)	Energy (Ryd)
GeKMg	7.26	7.22	21.2	-5803.059
SnKMg	7.61	7.57	18.392	-13963.145

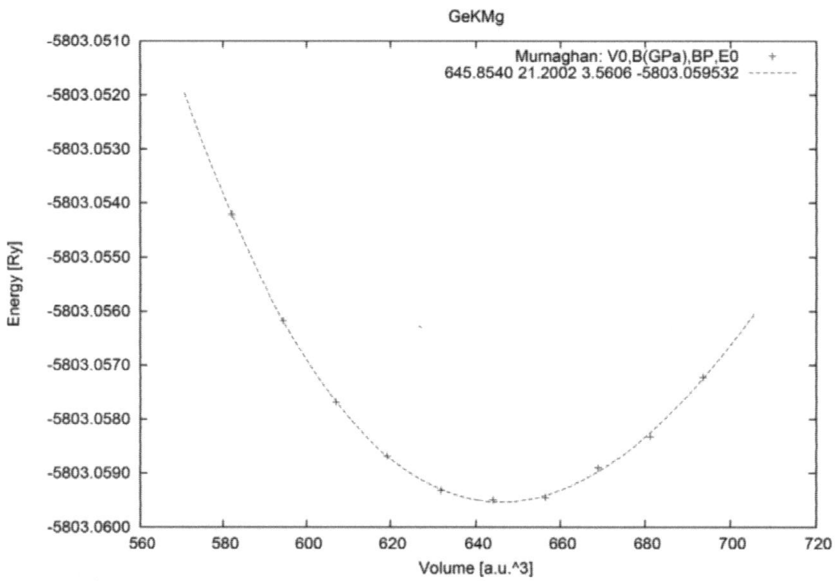

Figure 2. Total energy as a function of volume at equilibrium lattice constant for GeKMg in α-phase.

3.2 Electronic and magnetic properties.

Fig. 3 presents the spin polarized total density of states (DOS) of GeKMg and SnKMg at their respective equilibrium lattice constants in GGA and mBJ configurations. One can see that for GeKMg in both GGA and mBJ there is an energy gap around the Fermi level in the majority-spin (spin-up) channel while the minority-spin (spin-down) channel is strongly metallic, that is to say, it exhibit a true HM characteristic with 100% spin-polarization around the Fermi level. However for SnKMg structure the energy gap around the Fermi level in the majority-spin (spin-up) channel is not exactly around the Fermi level indicating that it is nearly half metallic in GGA. But for mBJ configurations Fermi level lies in between the energy band gap in majority-spin channel thereby showing a half metallic character.

It can easily be observed that the total DOS is mainly contributed by the Ge-p states while the d and p states of K and Mg respectively contributes to the DOS near the Fermi level. Fig. 3 also shows that, for GeKMg the energy gaps around the Fermi level in the majority-spin channel are about

0.32 eV(GGA) and 1.05 eV (mBJ) and for SnKMg 0.49 eV(GGA) and 1.1 eV(mBJ) which are similar to NiMnSb (0.5 eV) and CoMnSb (1 eV) [35]. Meanwhile, the HM gap [31–33], which is the minimum between the bottom energy of majority (minority) spin conduction bands with respect to the Fermi level and the absolute values of the top energy of majority (minority) spin valence bands, is 0.1 eV and 0.39 eV for GeKMg in mBJ and GGA respectively and 0.14 eV for SnKMg in mBJ. The different physical properties of GeKMg and SnKMg are given in Table 3.The calculated spin-projected partial DOS of GeKMg and SnKMg in the ferromagnetic state at the equilibrium lattice constant is shown in Fig. 3. The partial DOS of some states such as s and d states of Ge are omitted as their contribution is very small. Comparing the total and partial DOS with the band structure we can see that, for both majority- and minority-spin channels, the lowest band mainly originates from the Ge-p states and the bands around 6 eV are mainly formed by the K-d state. However, these states are located far from the Fermi level and the spinsplitting is very weak, so that their contribution to the form of half-metallicity in GeKMg and SnKMg is very small. The half-metallicity is mainly from the spin polarization of the Ge-p, K-d and Mg-p.

Fig. 5 shows the respective band structure plots of GeKMg and SnKMg in GGA and mBJ in both spin channels along with total DOS. In the valence region of the spin up and spin down channels, the bands are mostly due to the the Ge-p states in both GeKMg and SnKMg . The bands in the conduction regions are mostly due to K-d electrons states in both cases of GeKMg and SnKMg . For GeKMg in GGA and mBJ the width of the energy gap EG is the difference in energies of the maximum occupied band at the symmetry point Γ in the valence band and the minimum unoccupied band in the conduction region at a symmetry point Γ which gives a direct band gap . And in the case of SnKMg in GGA and mBJ the energy gap is also along the symmetry point Γ respectively giving a direct band gap. For GeKMg in GGA and mBJ the direct band gap obtained are 0.32 eV and 1.05 eV respectively. Similarly for SnKMg in GGA and mBJ a direct band gap obtained along the symmetry point Γ are 0.49 eV and 1.1 eV respectively . When mBJ potential calculations are employed it is observed that the valence bands shift downwards while the conduction band shifts upwards resulting in a wider band gap.

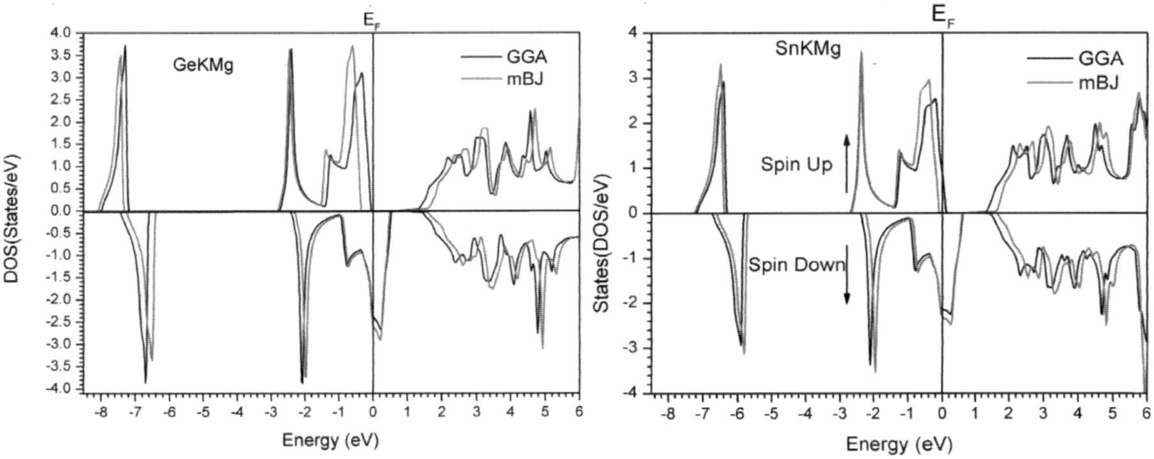

(a) (b)

Figure 3. Total DOS of GeKMg (a) and SnKMg (b) in GGA and mBJ at the lattice equilibrium constant.

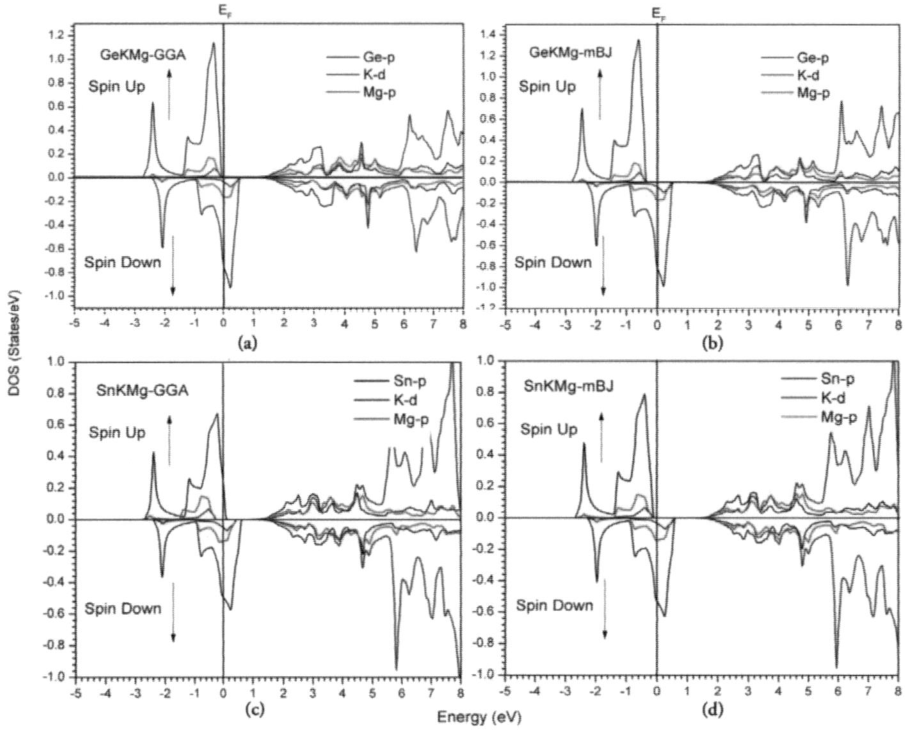

Figure 4. (a)Partial DOS of GeKMg in GGA (b) Partial DOS of GeKMg in mBJ (c) Partial DOS of SnKMg in GGA and Partial DOS of SnKMg in mBJ.

Table 3. Different physical properties of GeKMg and SnKMg at equilibrium lattice constant in the α-phase, VXC is the exchange correlation potential, VBM is the valence band maxima, CBM is the conduction band minima, EBG is the energy band gap, EHM is the half-metallic gap, M is magnetic moment, * gives the value of previous calculation [26].

Material	V_{XC} (eV)	V_{BM} (eV)	C_{BM} (eV)	E_{BG} (eV)	E_{HM} (eV)	M (μ_B)	M^* (μ_B)	Result	Result*
GeKMg	GGA	-0.1	0.22	0.32	0.10	0.99	1.00	HM	HM
	mBJ	-0.39	0.66	1.05	0.39	1.00	-	HM	-
SnKMg	GGA	0.07	0.56	0.49	-	0.90	0.81	Not HM	Not HM
	mBJ	-0.14	0.96	1.1	0.14	1.00	-	HM	-

Table 4. gives the calculated total and partial magnetic moments for GeKMg and SnKMg at the optimized equilibrium lattice constants. The calculated total magnetic moment per formula unit is 0.99 μ_B and 1 μ_B for GeKMg in GGA and mBJ respectively. While The calculated total magnetic moment per formula unit is 0.90 μ_B and 1 μ_B for SnKMg in GGA and mBJ respectively. It can be seen that the atomic spin moment mainly originates from the Ge and Sn atom in GeKMg and SnKMg respectively. Therefore from the observed magnetic moments we can conclude that GeKMg is a half metal in both GGA and mBJ while for SnKMg half metallicity is observed only in mBJ.

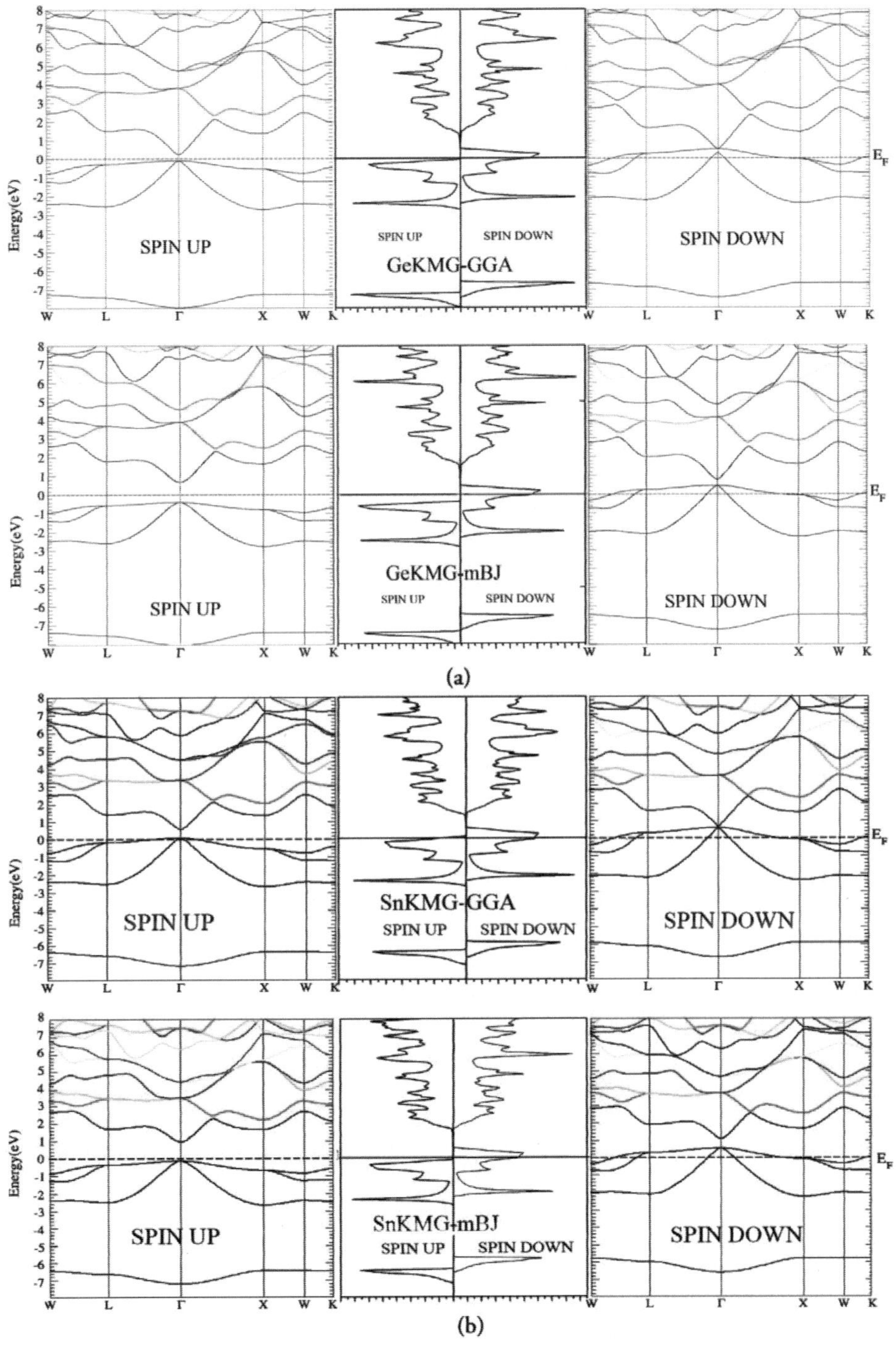

Figure 5. Band Structures of GeKMg (a) and SnKMg (b) in GGA and mBJ with DOS.

Table 4. Total magnetic moment M_{tot} (μB) and local magnetic moments (μB) of GeKMg and SnKMg in GGA and mBJ values. M_I (μB) is the magnetic moment of the interstitial region, M_{tot}*[26] is the previous calculation. V_{XC} is the exchange –correlation potential.

Compounds	V_{XC}	m_{Ge} (μB)	m_K (μB)	m_{Mg} (μB)	M_I (μB)	M_{tot} (μB)	M_{tot}* (μB)
GeKMg	GGA	0.41	0.03	0.03	0.51	0.99	1.00
	mBJ	0.45	0.03	0.01	0.45	1.00	-
SnKMg	GGA	0.25	0.03	0.03	0.57	0.90	0.81
	mBJ	0.29	0.02	0.03	0.64	1.00	-

4. Conclusion

To summarize, we have investigated the structural, electronic and magnetic properties GeKMg and SnKMg compounds with half-Heusler structure by using the full potentiacl linearised augmented-plane wave (FP-LAPW) code based on density functional theory implemented with the WIEN2k package. GGA approximation is first used to get a view of the electronic structures and mBJ potential methods are employed to get a better description of the energy band gaps. GeKMg showed half-metallicity with HM gaps of 0.1 eV(GGA) and 0.39 eV (mBJ) while SnKMg showed half-metallicity only in the mBJ configuration with HM gap of 0.14 eV. The calculated magnetic moment, 1 μ_B per formula unit for both GeKMg and SnKMg, mainly originates from the Ge (Sn) atom.

Acknowledgement

The author (T. Malsawmtluanga) would like to extend his sincere appreciation to Condensed Matter Theory Research Group, Physics Department, Mizoram University and especially to Benjamin Vanlalruata, a brilliant contributor to the group who recently passed away.

References

[1] De Groot CH, Buschow KH, De Boer FR. Magnetic properties of $R_2Co_{17-x}Al_x$ compounds (R= Ho, Dy, Y). Physica B: Condensed Matter. 1997 Mar 31;229(3):213-6.

[2] De Groot RA, Mueller FM, Van Engen PG, Buschow KH. New class of materials: half-metallic ferromagnets. Physical Review Letters. 1983 Jun 20;50(25):2024.

[3] Jaiganesh, G., Eithiraj, R. D., & Kalpana, G. (2010). Theoretical study of electronic, magnetic and structural properties of Mo and W based group V (N, P, As, Sb and Bi) compounds. Computational Materials Science, 49(1), 112-120.

[4] Şaşıoğlu E, Sandratskii LM, Bruno P, Galanakis I. Exchange interactions and temperature dependence of magnetization in half-metallic Heusler alloys. Physical review B. 2005 Nov 14;72(18):184415.

[5] Kübler J, Fecher GH, Felser C. Understanding the trend in the Curie temperatures of Co_2-based Heusler compounds: Ab-initio calculations. Physical Review B. 2007 Jul 10;76(2):024414..

[6] Galanakis I. Surface properties of the half-and full-Heusler alloys. Journal of Physics: Condensed Matter. 2002 Jun 14;14(25):6329.

[7] Buschow KH. Magneto-optical properties of alloys and intermetallic compounds. Handbook of Ferromagnetic Materials. 1988 Dec 31;4:493-595.

[8] Sattar MA, Rashid M, Hashmi MR, Rasool MN, Mahmood A, Ahmad SA. Spin-polarized calculations of structural, electronic and magnetic properties of Half Heusler alloys FeVX (X= Si, Ge, Sn) using Ab-initio method. Materials Science in Semiconductor Processing. 2016 Aug 15;51:48-54.

[9] Ghimire MP, Thapa RK, Rai DP, Sinha TP, Hu X. Half metallic ferromagnetism in tri-layered perovskites $Sr_4T_3O_{10}$ (T= Co, Rh). Journal of Applied Physics. 2015 Feb 14;117(6):063903.
[10] Akai H. Ferromagnetism and its stability in the diluted magnetic semiconductor (In, Mn) As. Physical Review Letters. 1998 Oct 5;81(14):3002.
[11] Al-zyadi JM, Gao GY, Yao KL. First-principles study on the (111) surface of half-metallic rocksalt VPo. Solid State Communications. 2012 Jul 31;152(14):1244-8.
[12] Ogawa T, Shirai M, Suzuki N, Kitagawa I. First-principles calculations of electronic structures of diluted magnetic semiconductors (Ga, Mn) As. Journal of magnetism and magnetic materials. 1999 May 1;196:428-9.
[13] Perdew JP, Burke K, Ernzerhof M. Generalized gradient approximation made simple. Physical review letters. 1996 Oct 28;77(18):3865.
[14] Sanvito S, Hill NA. Ground state of half-metallic zinc-blende MnAs. Physical Review B. 2000 Dec 15;62(23):15553.
[15] Tran F, Blaha P. Accurate band gaps of semiconductors and insulators with a semilocal exchange-correlation potential. Physical review letters. 2009 Jun 3;102(22):226401.
[16] Blaha P, Schwarz K, Madsen GK, Kvasnicka D, Luitz J. wien2k. An augmented plane wave+ local orbitals program for calculating crystal properties. 2001.
[17] Perdew JP, Burke K, Wang Y. Erratum: Generalized gradient approximation for the exchange-correlation hole of a many-electron system [Phys. Rev. B 54, 16 533 (1996)]. Physical Review B. 1998 Jun 15;57(23):14999.
[18] Murnaghan FD. The compressibility of media under extreme pressures. Proceedings of the National Academy of Sciences. 1944 Sep 15;30(9):244-7.
[19] Becke AD, Johnson ER. A simple effective potential for exchange. The Journal of chemical physics. 2006 Jun 14;124(22):221101.
[20] Tran F, Blaha P. Accurate band gaps of semiconductors and insulators with a semilocal exchange-correlation potential. Physical review letters. 2009 Jun 3;102(22):226401.
[21] Becke AD, Roussel MR. Exchange holes in inhomogeneous systems: A coordinate-space model. Physical Review A. 1989 Apr 1;39(8):3761.
[22] Malsawmtluanga T, Thapa RK. Ground state study of half-metallic ferromagnetism in the half-Heusler compound Gekca using density functional theory. In PROCEEDING OF INTERNATIONAL CONFERENCE ON RECENT TRENDS IN APPLIED PHYSICS AND MATERIAL SCIENCE: RAM 2013 2013 Jun 3 (Vol. 1536, No. 1, pp. 1047-1048). AIP Publishing.
[23] Blöchl PE, Jepsen O, Andersen OK. Improved tetrahedron method for Brillouin-zone integrations. Physical Review B. 1994 Jun 15;49(23):16223.
[24] Wolf SA, Awschalom DD, Buhrman RA, Daughton JM, Von Molnar S, Roukes ML, Chtchelkanova AY, Treger DM. Spintronics: a spin-based electronics vision for the future. Science. 2001 Nov 16;294(5546):1488-95.
[25] Žutić I, Fabian J, Sarma SD. Spintronics: Fundamentals and applications. Reviews of modern physics. 2004 Apr 23;76(2):323.
[26] Chen J, Gao GY, Yao KL, Song MH. Half-metallic ferromagnetism in the half-Heusler compounds GeKCa and SnKCa from first-principles calculations. Journal of Alloys and Compounds. 2011 Oct 20;509(42):10172-8.
[27] Rai DP, Ghimire MP, Thapa RK. Study of energy bands and magnetic properties of Co_2CrSi Heusler alloy. Bulletin of Materials Science. 2011 Oct 1;34(6):1219-22.
[28] Rai DP, Shankar A, Sandeep, Ghimire MP, Thapa RK. Electronic structure and magnetic properties of Co_2YZ (Y= Cr, Z= Al, Ga) type Heusler compounds: a first principle study. International Journal of Modern Physics B. 2012 Mar 30;26(08):1250071.

[29] Xiong L, Yi L, Gao GY. Search for half-metallic magnets with large half-metallic gaps in the quaternary Heusler alloys CoFeTiZ and CoFeVZ (Z= Al, Ga, Si, Ge, As, Sb). Journal of Magnetism and Magnetic Materials. 2014 Jun 30;360:98-103.

[30] Rozale H, Khetir M, Amar A, Lakdja A, Sayede A, Benhelal O. Ab-initio study of half-metallic ferromagnetism in the XCsSr (X= C, Si, Ge, and Sn) half-Heusler compounds. Superlattices and Microstructures. 2014 Oct 31;74:146-55.

[31] Gao GY, Yao KL, Şaşıoğlu E, Sandratskii LM, Liu ZL, Jiang JL. Half-metallic ferromagnetism in zinc-blende CaC, SrC, and BaC from first principles. Physical Review B. 2007 May 25;75(17):174442.

[32] Gao, G. Y., and K. L. Yao. "Half-metallic sp-electron ferromagnets in rocksalt structure: The case of SrC and BaC." Applied Physics Letters 91.8 (2007): 082512.

[33] Gao GY, Yao KL. Half-metallic sp-electron ferromagnets in rocksalt structure: The case of SrC and BaC. Applied Physics Letters. 2007 Aug 20;91(8):082512.

[34] Liu BG. Robust half-metallic ferromagnetism in zinc-blende CrSb. Physical Review B. 2003 May 30;67(17):172411.

[35] Galanakis I, Dederichs PH, Papanikolaou N. Slater-Pauling behavior and origin of the half-metallicity of the full-Heusler alloys. Physical Review B. 2002 Nov 18;66(17):174429.

[36] Picozzi S, Continenza A, Freeman AJ. Co 2 Mn X (X= Si, Ge, Sn) Heusler compounds: An ab initio study of their structural, electronic, and magnetic properties at zero and elevated pressure. Physical Review B. 2002 Sep 17;66(9):094421.

[37] Yang FY, Shang CH, Chien CL, Ambrose T, Krebs JJ, Prinz GA, Nikitenko VI, Gornakov VS, Shapiro AJ, Shull RD. Multistep magnetic switching in single-crystal (001) Co 2 MnGe films. Physical Review B. 2002 Apr 23;65(17):174410.

[38] Branford WR, Singh LJ, Barber ZH, Kohn A, Petford-Long AK, Van Roy W, Magnus F, Morrison K, Clowes SK, Bugoslavsky YV, Cohen LF. Temperature insensitivity of the spin-polarization in Co2MnSi films on GaAs (001). New Journal of Physics. 2007 Feb 28;9(2):42.

[39] Wurmehl S, Fecher GH, Kandpal HC, Ksenofontov V, Felser C, Lin HJ, Morais J. Geometric, electronic, and magnetic structure of Co 2 FeSi: Curie temperature and magnetic moment measurements and calculations. Physical Review B. 2005 Nov 30;72(18):184434.

[40] Jishi RA. Modified Becke-Johnson exchange potential: improved modeling of lead halides for solar cell applications. AIMS Materials Science. 2016 Jan 11;3(1):149-59.

[41] Jishi RA, Ta OB, Sharif AA. Modeling of lead halide perovskites for photovoltaic applications. The Journal of Physical Chemistry C. 2014 Nov 24;118(49):28344-9.

[42] Nanda BR, Dasgupta I. Electronic structure and magnetism in half-Heusler compounds. Journal of Physics: Condensed Matter. 2003 Oct 17;15(43):7307.

[43] Zhang M, Dai X, Hu H, Liu G, Cui Y, Liu Z, Chen J, Wang J, Wu G. Search for new half-metallic ferromagnets in semi-Heusler alloys NiCrM (M= P, As, Sb, S, Se and Te). Journal of Physics: Condensed Matter. 2003 Nov 7;15(46):7891.

[44] Zhang M, Liu ZH, Hu HN, Liu GD, Cui YT, Wu GH, Brück E, de Boer FR, Li YX. Half-metallic ferromagnetism in hypothetical semi-Heusler alloys NiVM „MÄP, As, Sb, S, Se, and Te…. Journal of applied physics. 2004 Jun 1;95(11).

[45] Sasioglu E, Sandratskii LM, Bruno P. Above-room-temperature ferromagnetism in half-metallic Heusler compounds NiCrP, NiCrSe, NiCrTe and NiVAs: A first-principles study. arXiv preprint cond-mat/0505299. 2005 May 11.

Investigation of manifestation of optical properties of butterfly wings with nanoscale zinc oxide incorporation

Swati N. Aideo, Dambarudhar Mohanta

[1]Dept. Of Physics, Nanoscience and Soft Matter Laboratory Tezpur university, Tezpur University, Napaam-784028, India

best@tezu.ernet.in; sswatinawami@gmail

Abstract. In this work, microstructural and optical characteristics nanoparticles of wings of Tailed Jay (Graphium Agamemnon) butterfly were studied before and after treating it in a precursor solution of zinc acetate and ethanol. We speculate that the butterfly scales are infiltrated with ZnO nanoparticles owing to reduction of Zinc hydroxide under ambient condition. The ZnO butterfly scales so produced were characterised using optical microscopy, UV-Vis reflectance spectroscopy, and electron microscopy etc. From the reflectance spectra, we could see that after treating it in the solution, optical properties vary. We anticipate that this change may be due to the formation of ZnO nanoparticles as well as the loss in periodicity due to the chemical treatments, which could be assessed from the SEM micrographs.

1. Introduction

Nature provides ample number of biological systems which display beautiful patterns and colours. The study of the microstructures in these systems gives hints about fabricating artificial photonic structures [1]. These biological systems provide novel platforms and templates so as to accommodate interesting inorganic materials. There are several biomaterials, such as bacteria [2] and fungal colonies [3], wood cells [4], diatoms [5], echinoid skeletal plates [6], pollen grains [7], eggshell membranes [8], human and dog's hair [9] and silk [10] which have been used for the biomimetic synthesis of a range of organized inorganic make ups that has potential in catalysis, magnetism, separation technology, electronics and photonics. Several research groups have worked on different novel methods to produce such inorganic materials using natural templates. In the recent years, butterfly wings as templates have gained enormous attention because of its complex yet extraordinary architecture. The wings of some of the butterflies contain natural photonic crystals which may be modified using semiconductor oxide nanoparticle incorporation. Gary Cook *et al.* have worked on the synthesis of silicify butterfly wings [11]. In his work, silicate minerals were deposited in the inner space of the wing scales and a heavy coating (100–150 nm thick) was produced on the surfaces by a chemical vapour deposition technique. Zhang et al. have worked on fabricating ZnO replica using butterfly wings and have claimed to fabricate three dimensional photonic crystals. But no prominent photonic properties are found to be observed [1, 12]. Zinc oxide is a wide band gap semiconductor oxide, which exhibits intense light emission features in the UV-Vis range of the spectra. There are also reports on the modification of the Photonic Band Gap (PBG) and the emission pattern of ZnO infiltrated PCs.

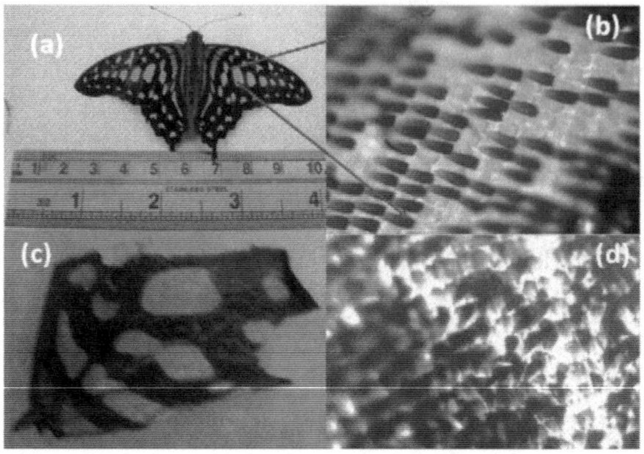

Figure 1. Digital image and optical micrograph of (a, b) untreated butterfly wing.

The present work highlights incorporation of ZnO nanoparticles in the forewing of Tailed Jay butterfly and subsequently, microstructural morphology and optical properties of the wings have been discussed.

2. Experimental: Materials And Methods

2.1. Material:

The material chosen for our work is a Tailed Jay butterfly (*Graphium Agamemnon*) of *papilionidae* family belonging to *G.agamemnon* species. The butterfly wing has black appearance with isolated green patterns distributed over it. We opted for the forewings of the butterfly, which measures typically of 4.3 mm and 2.3 mm in length and width respectively. The dimensions of the complete wing are approximately 8 mm and 5 mm in length and width respectively. The beautiful coloured bright patterns have motivated us to undertake this study. The digital image of the butterfly has been shown in Figure. 1.

2.2. Synthesis of ZnO incorporated butterfly wings:

The forewing of the butterfly is sectioned carefully for ZnO incorporation. Firstly, the wings were pretreated before growing ZnO particles on it [13]. The pretreatment process involves three steps: The wing slice was dipped in 5% HCl solution for 3 h. After washing carefully with deionised water, it was again dipped in a solution of 10% NaOH at 100°C for 4h. The wing slice was rinsed properly with deionised water before dipping it in a precursor solution prepared by dissolving zinc acetate $[Zn(O_2CCH_3)_2.2H_2O]$ in absolute ethanol in a closed vessel for 24 h. The wing was gently collected from the solution, washed with deionised water and dried in air, by placing the wing between two glass slides to keep the wing flat. The dried wings are then placed in an oven at 200°C for about 1 h. This would help growth of nanoparticles through reduction of zinc hydroxide available on the template.

3. Characterisation

The optical micrographs and digital images for the pure and ZnO grown forewings of the Tailed Jay butterfly are shown in Fig 1. The complex architecture of the butterfly wing was initially viewed through an inverted optical microscope (Leica DM IL LED) by employing a 20x objective. In this microscope, transmitted light is considered for optical imaging. The optical properties of the butterfly wings and ZnO grown wings are investigated through the UV-Vis-NIR spectrophotometer (2450 Shimadzu Co.) and an optical microscope. Further, scanning electron microscopy (SEM, JEOL, JSM 6390 LV) is employed for detailed microstructure analysis of the given specimen. Prior loading into the SEM chamber, the surface of the specimen was subjected to a few nm Pt coating to avoid charging effect during imaging. The images are captured at different magnifications to reveal microstructural details along with the chitinous elements that make up the scale structures.

4. Results And Discussions

4.1. Imaging through optical microscopy

The digital image of the specimen is shown in figure 1 (a). Figure 1(b) depicts the optical micrograph of the green part of the butterfly wing. Figure 2 (c, d) presents the digital image and the optical image for the part of the wing which was dipped in the precursor solution. The properly arranged cover and ground scales are found to be disrupted and the periodicity is found to be lost due to chemical treatment.

4.2. Measurement through UV-Vis-NIR reflectance spectroscopy:

The optical properties of the pure butterfly wings and ZnO infiltrated butterfly wings are studied through UV-Vis spectroscopy. The reflectance spectra of the butterfly wings before and after ZnO inclusion are as shown in figure 2.

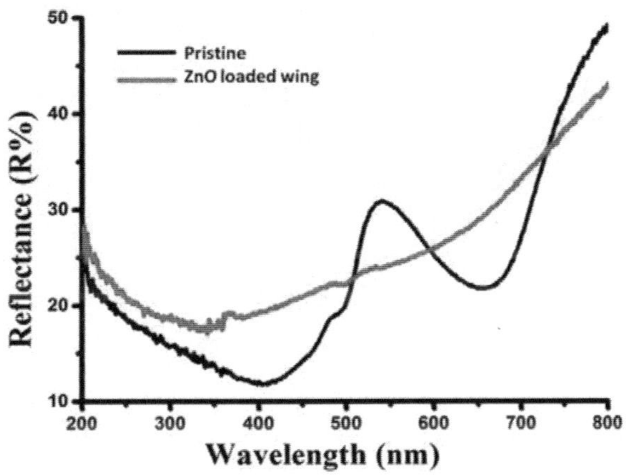

Figure 2. Reflectance spectra for the pristine butterfly wings and after ZnO loading.

The reflectance spectra of the untreated as well as ZnO loaded butterfly wings exhibit a small peak at ~ 485nm and a distinct peak at ~ 539 nm. The green coloured patterns distributed over the highly pigmented black surface corresponds to 539 nm peak and this may be as a result of the unique microstructure in the wing scales. After ZnO is grown over the wing surface, there is an alteration in the reflectance spectra with featureless characteristics and a linear rise towards the higher wavelength regime. This change can be attributed to the change in surface morphology due to ZnO loading as there is a strong probability that nanoparticle incorporation might fill in the spaces between cross ribs and ribs, leading to change in refractive index and hence in the optical properties [1]. Non observation of peaks can also be due to the loss of periodicity, as we observe from the OM image. We anticipated that ZnO nanoparticles would be decorated on the wing, but formation might not be perfect, as a result, leaving with Zn^{2+} ions. Also, there is a probability that ZnO nanoparticles might have formed but some of the unreacted sites may be left with only Zn^{2+} ions.

4.3. Microstructural analysis of surface structure:

The SEM micrographs for the untreated and ZnO loaded butterfly wings are shown in figure 3.

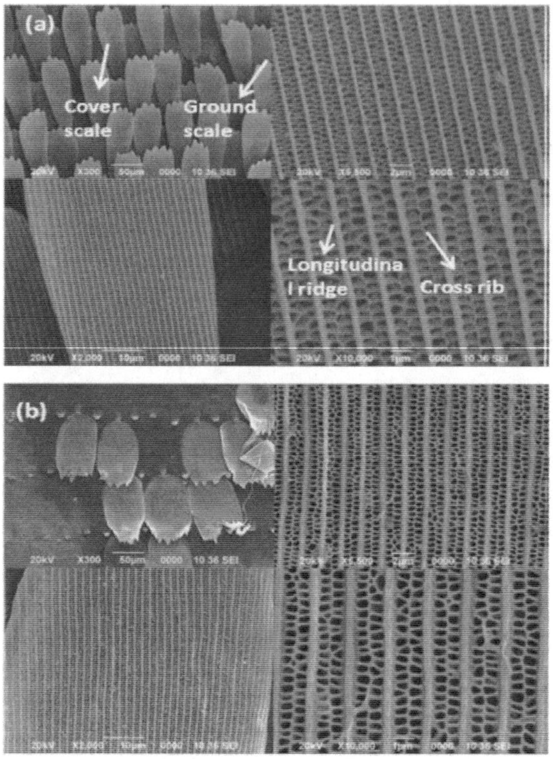

Figure 3. SEM micrographs for the (a) pristine butterfly wings (b) ZnO loaded wings.

From the SEM micrographs, we see the presence of cover scales and ground scales distributed uniformly all over the wing surface. The ground scales lie directly above the wing lamina whereas the cover scales overlay these [14]. The length and width of the cover scales are 131.80 μm and 44 μm respectively. Further magnification reveals the presence of different components that make up the scales. The longitudinal ridges are distributed parallel all along the wing surface. The cross ribs are interconnected in such a manner so as to form a gyroid type structure and are cross linked to each other [15]. The approximate dimensions of the windows between the ribs and cross ribs are 568 nm and 409 nm in length and width; respectively. The width of the longitudinal ridges is approximately 340 nm. When the treated wing scales undergo heat treatment, the cover and ground scales are seen to be broken and changes are also observed in the dimensions of the scales and the scale components. The gyroid type structure takes more of honeycomb like structure after ZnO incorporation. The nanopores become quite clear and they shrink to approximately 509 nm and 314 nm in length and width respectively. Although precise chemical composition of the scales is not known, all insect cuticles primarily contain chitin layer and other protein groups like resilin or scleretin. Surface functional groups on the chitin layer such as hydroxyl and amino groups may act as centres and bring Zn^{2+} ions close through hydrogen bond formation [16]. Table below shows the change in scale parameters before and after ZnO loading and the calculated shrinkage factor.

4.4. Correlation between surface morphology and optical properties

The reflectance properties are closely related to the surface morphology of the incident material. As a result of ZnO incorporation on the wing surface, wing scales are found to be deformed, and the dimension of the scale components varies. This alteration will have a strong effect on the optical properties. The distinct green peak that is found in pure butterfly wings is found to disappear in the ZnO grown wings. The arrangements of the ridges, cross ribs on the cover scales of the wing are in such manners that form a multilayer structure, with the dimensions of the nanopores approximately of 568 nm, which may be attributed to the green colour of the wing. Previous studies have reported that the cover scales are mainly responsible for the reflected colour, whereas the ground scales contains the pigment melanin, where it absorbs almost all of the transmitted light [14], thus appearing black.

Table 1: Scale parameters calculated from SEM image.

Scale parameter	**Before ZnO loading (L1) μm**	**After ZnO loading (L2) μm**	**Shrinkage factor [1- (L2/L1)] %**
Length of cover scale	131.78	116	11.9
Width of cover scale	44	63.8	
Width of longitudinal ridge	0.34	0.33	2.9
Spacing between longitudinal ridges	1.5	1.37	8.66
Spacing between two cross ribs	0.636	0.40	37

5. Conclusions

In this work, we made an attempt to understand the effect of ZnO incorporation on the wings of Tailed jay butterfly belonging to *Papillionidae* family and *G. agamemnon* species. The optical properties studied through UV-Vis reflectance spectroscopy for the pure as well as treated samples gave us an idea regarding the alteration in optical properties as a result of nanoparticle incorporation and also due to loss in periodicity. The scale components are found to shrink due to chemical and heat treatment, with different percentages, with a highest of 37% for the spacing between the cross ribs. This decrease in the dimension of the nanopores as a result of ZnO incorporation contributes to the change in the reflectance spectra. Thus, this study shows how structure and optical properties change as a result of nanoparticle inclusion.

Acknowledgments

The authors thank Mr. R. Boruah for assisting in scanning electron microscopy imaging. We also extend our sincere thanks to colleagues for the valuable suggestions.

References

[1] Zhang Wang, Zhang Di, Fan Tongxiang, Jian Ding, Gu Jiajun, Guo Qixin and Ogawa Hiroshi 2006 *Bioinsp. Biomim.* **1** 89

[2] Davis S A, Burkett S L, Mendelson N H and Mann S 1997 *Nature* **385** 420

[3] Li Z, Chung S W, Nam J M, Ginger D S, Mirkin C A 2003 *Angew Chem Int Ed Engl.* **20** 2306
[4] Shin Y, Liu J, Chang J H, Z Nie and Exarhos G J 2001 *Adv. Mater.* **13** 728
[5] Anderson M W, Holmes S M, Hanif N and Cundy C S 2000 *Angew. Chem. Int. Ed. Engl.* **39** 2707
[6] Meldruma Fiona C and Seshadri Ram 2000 *Chem. Commun.* 29
[7] Hall S R, Bolger H, Mann S 2003 *Chem Commun (Camb).* **22** 2784
[8] Yang D, Qi L M and Ma J M 2002 *Adv. Mater* **14** 1543
[9] Kim Y 2003 *Biomacromolecules* **4** 908
[10] He J and Kunitake T 2004 *Chem. Mater.* **16** 2656
[11] Cook Gary, Timms Peter L, Göltner-Spickermann Christine 2003 *Angew. Chem. Int. Ed. Engl.* **42** 5527
[12] Zhang J Z, Gu Z-Z, Chen H-H, Fujishima A and Sato O 2006 *J. Nanosci. Nanotechnol.* **6** 1173
[13] Zhang W, Zhang D, Fan T X, Ding J, Guo Q X and Ogawa H 2006 *Micropor. Mesopor. Mat.* **92** 227
[14] Giraldo M A, Stavenga D G 2016 *J Comp Physiol. A* **202** 381
[15] Yoshioka S, Fujita H, Kinoshita S and Matsuhana B 2013 *J. R. Soc. Interface* **11** 1
[16] Majeti N V and Ravi K 2000 *React. Funct. Polym.* **46** 1

Journal of Physics: Conference Series **765** (2016) 012020 doi:10.1088/1742-6596/765/1/012020

Structural and Dielectric Properties of Ionic Liquid Doped Metal Organic Framework based Polymer Electrolyte Nanocomposites

Rituraj Dutta[1], Ashok Kumar[1]

[1]Department of Physics, Tezpur University, Napaam, Tezpur, Assam, India
Corresponding author's e-mail: ask@tezu.ernet.in

Abstract. Metal Organic Frameworks (MOFs) are mesoporous materials that can be treated as potential hosts for trapping guest molecules in their pores. Ion conduction and phase behavior dynamics of Ionic Liquids (ILs) can be controlled by tunable interactions of MOFs with the ILs. MOFs incorporated with ionic liquid can be dispersed in the polymers to synthesize polymer electrolyte nanocomposites with high ionic conductivity, electrochemical and thermal stability for applications in energy storage and conversion devices such as rechargeable Li-ion batteries. In the present work we have synthesized Cu-based MOF [Cu_3(1,3,5-benzene tricarboxylate)$_2$(H_2O)] incorporated with the ionic liquid 1-Butyl-3-methylimidazolium bromide at different weight ratios of MOF and IL. The synthesized MOF-IL composites are dispersed in Poly (ethylene oxide) (PEO). Frequency dependent behavior of permittivity and dielectric loss of the nanocomposites depict the non-Debye dielectric relaxation mechanism. The room temperature Nyquist plots reveal decreasing bulk resistance upto 189 Ω with optimum ionic conductivity of 1.3×10^{-3} S cm^{-1} at maximum doping concentration of IL in the nanocomposite system.

1. Introduction

Metal Organic Frameworks (MOFs) are uniformly sized microporous group of hybrid materials containing an infinite network of inorganic metal clusters coordinated with organic linkers through metal ligand coordination bonds [1]. Owing to their high surface area and large pore volume, MOFs can be used in different applications such as gas adsorption, purification, separation, catalysis and ion exchange [2]. MOFs are highly designable and can be treated as host materials for small guest molecules to enhance their electrochemical properties. Ionic Liquids (ILs) are molten salts consisting of asymmetric organic cations and symmetric inorganic anions that can undergo unlimited structural variations and have a melting point below 100°C [3]. In recent years, researchers are paying much attention towards ionic liquids because of their significant properties such as low vapour pressure, high viscosity, good solvability for both organic and inorganic compounds, high thermal and electrochemical stability and high ionic conductivity. Due to their unique properties ionic liquids have potential applications in electrochemical devices, enzymatic catalysis, sensing and solar energy conversion [4]. MOFs can be used as potential host materials for controlling the phase behavior dynamics of ILs via the nanosizing of ILs and tunable interactions of MOFs with the ILs [5].

Kazuyuki Fujie et al. [6] investigated the phase behavior and ionic conductivity of lithium ion-doped ionic liquid (IL) EMI-TFSA incorporated into ZIF-8 MOF. The bulk (EMI0.8Li0.2) TFSA showed two orders of magnitude higher ionic conductivity than that of (EMI0.8Li0.2) TFSA in the micropores. The composite polymer electrolytes based on MOF can also be a good approach to enhance the performance parameters in energy storage and conversion devices. N. Angulakshmi et al. [7] prepared Magnesium-benzene tricarboxylate metal organic framework (Mg-BTC MOF)-incorporated composite polymer electrolytes (CPE) composed of poly(ethylene oxide) (PEO) and lithium bistrifluoromethane sulfonylimide (LiTFSI). The ionic conductivity of CPE incorporated with Mg-BTC MOF was enhanced up to two orders of magnitude even at 0°C with improved thermal stability and compatibility. The discharge capacity of the lithium polymer cell composed of Li/CPE/$LiFePO_4$ was found to be 110 mAh g^{-1} with a current rate of 1-C at 70°C. Claudio Gerbaldi et al. [8] reported that the incorporation of aluminium-based MOF as the filler to poly(ethylene oxide)-based nanocomposite polymer electrolyte (NCPE) membranes can enhance the ionic conductivity of the system up to two orders of magnitude with high stability and excellent cycling profile at moderate temperature.

In the present work a Cu based MOF [Cu_3(1,3,5-benzene tricarboxylate)$_2$(H_2O)], (popularly known as MOF-199) has been incorporated with ionic liquid (1-butyl-3 methylimidazolium bromide) with different concentrations and dispersed in polymer Poly(ethylene oxide) matrix with a view to enhance the structural and electrochemical properties of the nanocomposite system.

2. Experimental

2.1. Materials

Copper (II) acetate monohydrate was purchased from SD Fine-Chem Limited, india. Benzene tricarboxylic acid, N-N-Dimethyl Formamide, Triethylamine, Ethanol and Acetonitrile were purchased from Merck Specialities Pvt. Ltd., India. 1-butyl-3-methylimidazolium bromide and Poly (ethylene oxide) were purchased from Sigma Aldrich.

2.2. Preparation of Cu based Metal Organic Framework

Cu based Metal Organic Framework has been prepared by solvothermal technique. Two separate solutions in 1:1:1 ratio of DMF, ethanol and distilled water (4 ml each), have been prepared. 0.86 g of copper acetate monohydrate and 0.5 g of benzenetricarboxylic acid were dissolved separately into the prepared solutions. The two samples were mixed and stirred for 30 minutes to mix them homogeneously. 0.5 ml of triethylamine was added drop wise to the magnetically stirred solution [9]. The solution was kept stirring for 24 h, followed by filtering and washing with 20 ml of DMF. The sample was kept under vacuum drying for 3 – 4 h at 373K and 1.5 g bluish green colored product yield was obtained.

2.3. Preparation of IL doped MOF based polymer electrolyte nanocomposite

MOF-IL composites were prepared by direct mixing using mortar and pestle at 30%, 40% and 50% of doping concentrations of IL. MOF-IL Polymer electrolyte nanocomposites were prepared by solution casting method. 1g of poly (ethylene oxide) (PEO) was dissolved in 20 ml of acetonitrile and stirred for 4 h at 50° C. Then 0.5 g of each MOF-IL composite of different doping concentration was dispersed separately into the polymer solution and allowed mixing by ultrasonication for 2 h. Then each of the solutions were cast on petri- dishes and dried at room temperature for 1-2 days to get flexible bluish white colored films having thickness 30-40 μm.

2.4. Analysis and Characterization Techniques

The XRD patterns of pristine MOF-polymer films and doped ones were obtained by Advanced D8 X-ray diffractometer (Bruker AXS, Germany). Thermogravimetric analyzer (Perkin Elmer STA 6000) was used for analyzing the thermal stability of the pristine and IL doped MOF-polymer films. The surface morphology of MOF-IL composite was observed by a scanning electron microscope (SEM, Jeol 6390 LV). The ac impedance studies of MOF-IL polymer nanocomposite films were investigated by Hioki 3532-50 LCR HiTester in the frequency range from 42 Hz to 5 MHz.

3. Results and Discussion

Figure 1 shows the XRD patterns of pristine and IL doped MOF-PEO nanocomposites. The peaks at (120) and (112) planes correspond to the diffraction pattern of PEO. The peaks at (222) and (400) reflect the formation of MOF-199. Due to the incorporation of ionic liquid into the MOF-polymer system, the intensities of the XRD peaks decrease substantially. Due to the interaction of $BMIM^+$ and oxygen atom of PEO chain, the degree of crystallinity is decreased with increasing concentration of ionic liquid [10]. The crystallite size (L) has been calculated using Scherrer equation:

$$L = K\lambda / \beta \cos\theta \qquad (1)$$

where K is the crystallite shape constant taken to be 0.89, λ is the X-ray wavelength (1.5406 Å), β is the value of full width at half maximum of diffraction peaks at angle 2θ. The crystallite size of the nanocomposite system is found to decrease from 4.46 Å to 3.63 Å with increasing concentration of IL from 30% to 50%.

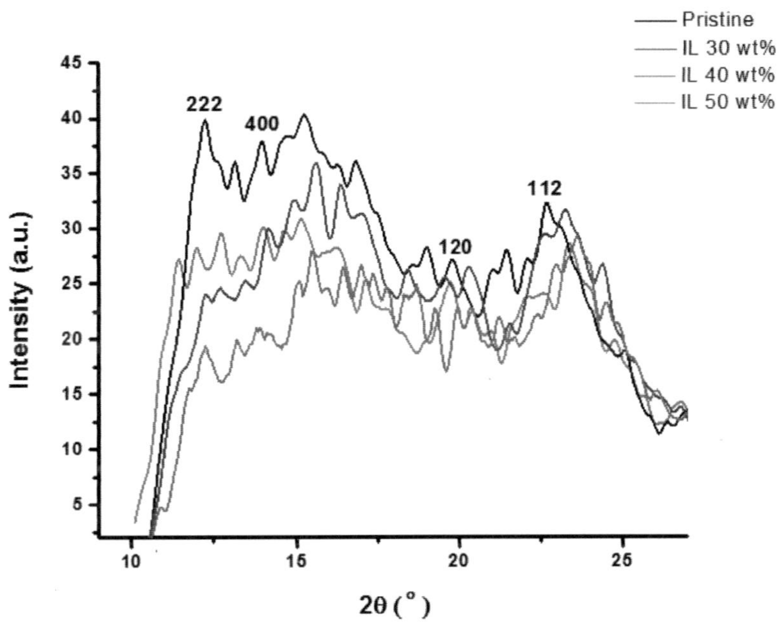

Figure 1. X-Ray Diffraction pattern of pristine and ionic liquid doped MOF-PEO nanocomposites.

Thermal stability of pristine and IL doped MOF-PEO nanocomposites is shown by thermogravimetric plots in Figure 2. Reduction in thermal stability is observed due to the enhancement in amorphicity with increasing doping concentration of ionic liquid in the MOF-PEO nanocomposites. The pristine sample shows thermal stability upto 310°C while in case of IL doped ones thermal stability is decreased upto 214°C.

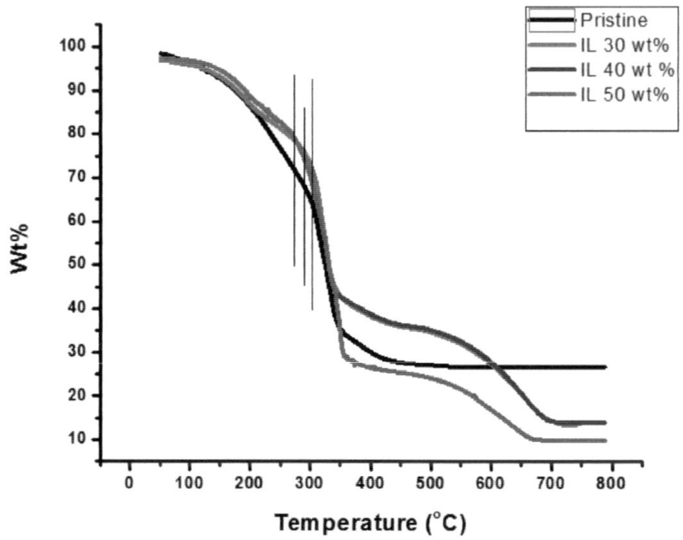

Figure 2. Thermogravimetric analysis of pristine and IL doped MOF-PEO nanocomposites.

The Scanning Electron Microscopy image of IL doped MOF is represented in Figure 3 shows crystalline morphology of the nanocomposite system.

Figure 3. SEM micrograph of IL doped MOF nanocomposite.

Figure 4 represents the frequency dependence of the real (ε') part of the dielectric function for the pristine and IL doped MOF polymer nanocomposite at room temperature. This can be attributed to the non Debye type behavior of accumulation of space charge region at the interface of the electrolyte. The real part of the dielectric function can be calculated using the formula:

$$\varepsilon' = \frac{Cd}{\varepsilon_0 A} \tag{2}$$

where, d is the thickness of the polymer film, C is the measured capacitance value, A is the area of the film and ε_0 is the permittivity of free space [11]. As crystallinity of the samples decreases upon increasing doping concentration of the ionic liquid, segmental motion of the polymer chain increases and that can enhance the asymmetry in relaxation process.

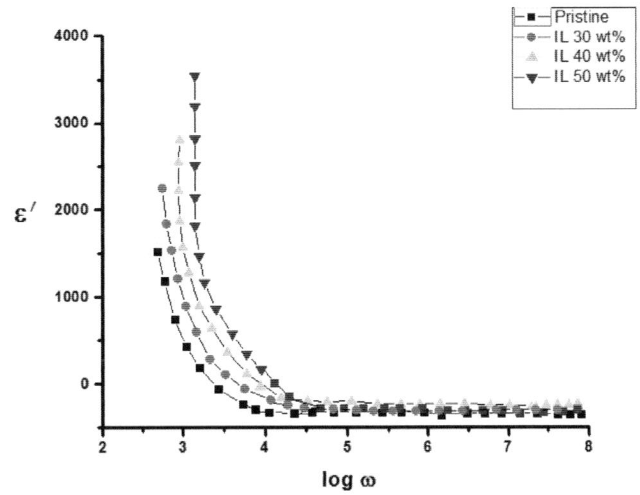

Figure 4. Variation of real part of permittivity (ε') with angular frequency.

The dielectric loss with the room temperature variation of frequency for pristine and IL doped MOF-PEO nanocomposites is depicted in Figure 5. Shifting of tan δ peaks towards high frequency side with increasing IL doping is observed. This is due to the enhancement of hopping frequency of the charge carriers with the increase in uptake of ionic liquid in the porous nanocomposite system.

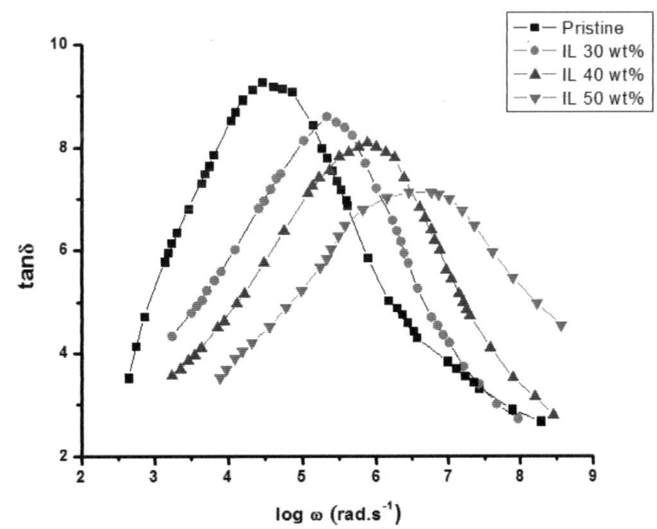

Figure 5. tan δ vs. angular frequency plots for pristine and IL doped MOF-PEO nanocomposites.

The room temperature Nyquist impedance plots of the pristine and IL doped MOF-PEO nanocomposites are shown in Figure 6. The plots contain semicircular arcs at the high-frequency that are due to the parallel combination of capacitances (C_g) and bulk resistances (R_b). The bulk resistance represents the movement of ions and the capacitance represents the dielectric polarization of polymer nanocomposites [12].

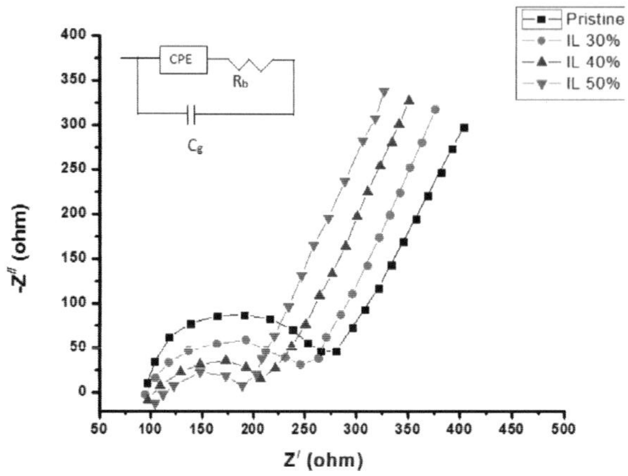

Figure 6. Room temperature Nyquist plots for pristine and IL doped MOF-PEO nanocomposites.

At the low-frequency region spikes are observed due to the formation of electric double layer. In inset of Figure 6 the equivalent circuit is shown having R_b in parallel with C_g in series with the constant phase element (CPE) [13]. Enhancement in ionic conductivity is observed as the bulk resistances of the nanocomposites are found to be decreased from 274 Ω to 189 Ω with increasing doping concentrations of ionic liquid. At maximum doping concentration of 50% of IL, ionic conductivity of 1.3×10^{-3} S cm^{-1} has been obtained.

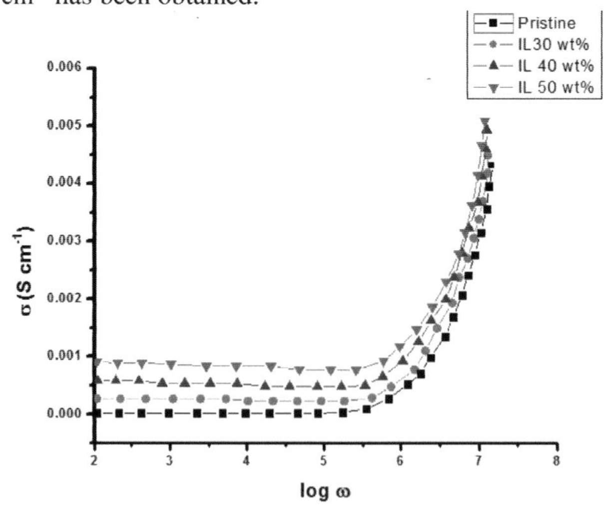

Figure 7. Room temperature variation of ac conductivity with frequency of pristine and IL doped MOF-PEO nanocomposites.

The room temperature variation of ac conductivity with frequency of pristine and IL doped MOF polymer electrolyte nanocomposites is shown in Figure 7.The dispersive behavior of ac conductivity at higher frequency regions is due to the non random forward-backward hopping of charge carriers from one coordination site to the others [14]. With increasing doping concentration of ionic liquid the segmental motion of the polymer chain enhances leading to long range hopping of ions [15].

4. Conclusion

Structural and dielectric properties of Cu based MOF polymer electrolyte nanocomposites have been investigated for different doping concentrations of ionic liquid. XRD patterns reveal the reduction in crystallinity as well as crystallite size from 4.46 Å to 3.63 Å upon IL doping. Permittivity spectra depict the dielectric relaxation mechanism in the IL doped nanocomposite systems which is dominated by hopping of ions. Optimum ionic conductivity of 1.3×10^{-3} S cm^{-1} has been obtained at maximum doping concentration of IL. Thermogravimetric analysis shows the reduction in thermal stability due to enhancement in amorphicity with increasing doping concentration of ionic liquid into the pores of MOF. The ac conductivity studies confirm the hopping of ions to new coordination sites to intensify ionic conductivity.

Acknowledgement

The authors gratefully acknowledge DST (INSPIRE), Govt. of India for providing financial support.

References

[1] Batten S R , Champness N R , Chen X M , Garcia Martinez J, Kitagawa S, Öhrström L, O'Keeffe M and Suh M P, Reedijk J 2013 *Pure and Applied Chemistry* **85** 1715-24
[2] Li Bin, Wen Hui Min, Zhou Wei and Chen Banglin 2014 *J. Phys. Chem. Lett.* **5** 3468–79
[3] Endres F and Zein El Abedin S 2006 *Phys. Chem. Chem. Phys.* **8**, 2101–16
[4] Plechkova N V and Seddon K R 2008 *Chem. Soc. Rev.* **37**, 123-150
[5] Fujie Kazuyuki, Otsubo Kazuya, Ikeda Ryuichi, Yamada Teppei and Kitagaw Hiroshi 2015 *Chem. Sci.* **6** 4306-10
[6] Fujie Kazuyuki, Yamada Teppei, Ikeda Ryuichi and Kitagawa Hiroshi 2014 *Angew. Chem. Int. Ed.* **53** 11302 – 05
[7] Angulakshmi N, Kumar R Senthil, Kulandainathan M Anbu and Manuel Stephan A 2014 *J. Phys. Chem. C* **118** 24240-47
[8] Gerbaldi Claudio, Nair Jijeesh R , Kulandainathan M Anbu, Kumar R Senthil, Ferrara Chiara, Mustarelli Piercarlo and Stephan Arul Manuel 2014 *J. Mater. Chem. A* **2** 9948-54
[9] Brinda L, Rajan K S and Ryappan John Bosco Balaguru 2012 *Journal of Applied Sciences* **12** 1778-80
[10] Polu A R and Rhee H W 2016 *International Journal of Hydrogen Energy* in press
[11] Nath A K and Kumar A 2014 *Ionics* **20** 1711-21
[12] Takeuchi I , Asaka K , Kiyohara K , Sugino T, Mukai K and Randriamahazaka H 2010 *J Phys Chem C* **114** 14627-34
[13] Khaled K F and Heckerman N 2003 *Electrochim Acta* **48** 2715-23
[14] Roling B , Martiny C and Funke K 1999 *J Non-Cryst Solids* **249** 201-09
[15] Singh P K, Kim K W and Rhee H W, 2008 *Electrochem Commun* **10** 1769-72

Effect of ZnS nanoparticles on the photoluminescence of Sm^{3+} ions in methanol

D Kakoti[1], N Rajkonwar[1], N Dehingia[1], A Boruah[1], P Gogoi[2] and *P Dutta[1]

[1]Laser and Spectroscopy Laboratory, Department of Physics, Dibrugarh University 786004, Dibrugarh, Assam, India.
[2]Dibrugarh University Institute of Engineering and Technology, 786004 Dibrugarh, Assam, India.

*pankaj@dibru.ac.in

Abstract: ZnS nanoparticles co-doped with Sm^{3+} ions were prepared in methanol medium for fixed Sm^{3+} and varying ZnS concentrations. Enhancements in absorption as well as photoluminescence efficiency of the co-doped samples were observed. This enhanced efficiency is attributed to the effective increase in oscillator strengths of the Sm^{3+} transitions because of the addition of ZnS nanoparticles.

Keywords: Semiconductor nanoparticles, rare earth, photoluminescence, energy transfer.

1. Introduction

The photoluminescence (PL) properties of rare-earth (RE) elements have been a vast area of research for a few decades. An attractive feature of REs is their sharp and stable emission which depends strictly on the tri-positive RE ion and to some extent on the environment or host [1]. The Sm^{3+} ion is one of the important RE ions which exhibits intense PL in the visible region. It finds wide applications in high-density optical storage, under sea communications, colour displays and visible solid-state lasers [2]. The application domains of Sm^{3+} mainly depend on its emission characteristics. Hence there is a need of enhancing the emission efficiency of Sm^{3+} in different hosts. One potential way of enhancement of the emission efficiency is by sensitization of the Sm^{3+} ions with different entities viz. nanoparticles (NP), ligands, other REs etc. Semiconductor NPs possess very good potential in the area of sensitized luminescence because the band gap energy of the semiconductor NP is of the order of the emission energy of the RE ions. Also, since the absorption cross section of the semiconductor NPs is quite high in comparison to the RE ions, they can absorb excitation energy from the source and can sensitize the RE ions by the recombination of the electron hole pairs [3, 4]. ZnS NPs have been extensively studied due to their outstanding optical properties. These are promising materials for absorption and emission in UV and visible (VIS) regions [5]. Nanosized ZnS exhibits discrete electron energy levels with high oscillator strength and strong luminescence due to the well known quantum confinement effects. Hence it bears potential as effective sensitizer for tri-positive RE ions [6]. The PL enhancement of Eu^{3+} due to the energy transfer from ZnS to Eu^{3+} in ZnS:Eu^{3+} nanocrystal has been reported by Sun *et al.* [7]. Chen et al. have studied the energy transfer from ZnS to Tb^{3+} with the help of absorption, excitation and PL of ZnS/ Tb^{3+} in water, isooctane microemulsion and sodium bis sulfosuccinate [8]. ZnS doped with Sm^{3+} is reported to be a good red emitter in thin film electroluminescence devices by T. Tohda *et al* [9].

In this present work we report the effect of ZnS NPs on the enhancement of PL efficiency of Sm^{3+} ions in methanol.

2. Experimental

At first the ZnS NPs were prepared in DMSO by mixing zinc nitrate hexahydrate ($Zn(NO_3)_2.6H_2O$, Loba Chemie, 96.103%) in dimethyl sulfoxide (DMSO, SRL 99%). The mixture was then heated to 60°C and kept at that temperature for five hours. The resultant solution became whitish signifying the formation of ZnS NPs. Stock solution of Samarium was prepared by dissolving Samarium nitrate hexahydrate ($Sm(NO_3)_3.6H_2O$, Sigma Aldrich, 99.9%) in 2 ml methanol and magnetically stirring for 15 minutes. The two solutions were then mixed and further stirred magnetically for 30 minutes. ZnS-Sm^{3+} doped samples were prepared for three different concentrations of ZnS. The Sm^{3+} concentration for all the samples were taken as 0.05M. The prepared samples were labeled as S1, S2, S3 and S4 where S1 contains only Sm^{3+} in methanol whereas S2, S3 and S4 contain both Sm^{3+} and ZnS in methanol with molar concentration of ZnS as 0.05, 0.15, 0.5 respectively.

The absorption spectra for the samples were recorded with a Shimadzu UV- 2600 UV-Vis spectrophotometer by using methanol as reference. PL and excitation spectra were recorded with Horiba Jobin Yvon Fluoromax-4P spectrofluorometer. All the recordings were done at room temperature.

3. Result and Discussion

3.1. Absorption

Fig. 1 shows the absorption spectra in the UV-VIS region of the Sm^{3+} co-doped with and without ZnS in Methanol. The Sm^{3+} absorption peaks are observed at 287 nm ($^6H_{5/2}\rightarrow(^4I,\ ^4F)_{9/2}$), 345 nm ($^6H_{5/2}\rightarrow{}^4D_{7/2}$), 363 nm ($^6H_{5/2}\rightarrow{}^4D_{3/2}$), 376 nm ($^6H_{5/2}\rightarrow{}^6P_{7/2}$), 390 nm ($^6H_{5/2}\rightarrow{}^4L_{15/2}$), 403 nm ($^6H_{5/2}\rightarrow{}^6P_{3/2}$), 417 nm ($^6H_{5/2}\rightarrow(^6P,\ ^4P)_{5/2}$), 440 nm ($^6H_{5/2}\rightarrow{}^4G_{9/2}$), 464 nm ($^6H_{5/2}\rightarrow{}^4I_{9/2}$), and 478 nm ($^6H_{5/2}\rightarrow{}^4I_{11/2}$) [10- 12,14]. Except the peak at 403 nm, all the peaks in the range 340-480 nm have very small intensity and expanded in the inset of Fig. 1. The 310 nm is the absorption peak of ZnS due to the electron-hole confinement in a small volume as a result of quantum size effect. The band-gap of the semiconductor is larger than the bulk semiconductor (335 nm) which confirms the particles are in the nano range [6]. The increase in intensity of the 310 nm absorption peak with the increase in concentration of ZnS, indicates the increase of particle number. The slight red shifting of this absorption peak might indicate the increase in particle size with the increase in ZnS concentration.

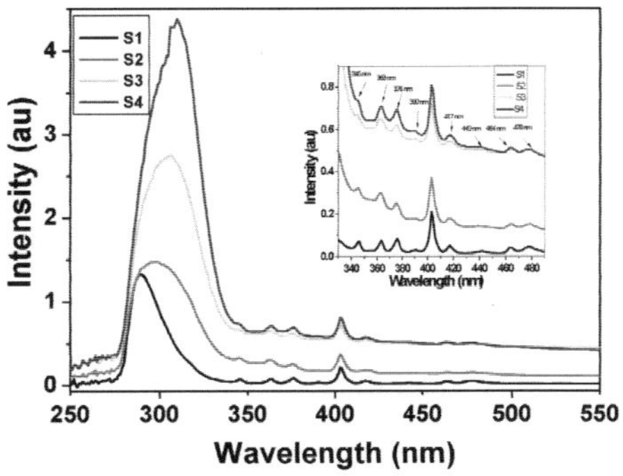

Figure 1: Absorption spectra for singly doped Sm^{3+} and ZnS co-doped Sm^{3+} samples. Inset shows the expanded view of absorption spectra showing Sm^{3+} transitions in the range 340-480 nm.

3.2. Excitation and Photoluminescence:

Fig. 2 shows the excitation spectra for S2, S4 and singly doped ZnS samples recorded in the range of 300-550 nm for emission wavelength fixed at 596 nm. For the Sm^{3+} doped samples, the spectra show maximum intensity for the excitation peak at 403 nm corresponding to $^6H_{5/2}\rightarrow{}^6P_{3/2}$ transition of Sm^{3+}, because of the highest oscillator strength of the transition [2, 14, 15]. This fact is also revealed by the absorption spectra in Fig.1. The excitation spectrum of ZnS singly doped sample also shows a broad absorption in the range of 325 to 500nm which considerably overlaps with the absorption maxima of samples S2 and S4. Another excitation spectrum for singly doped ZnS sample was recorded for the emission wavelength fixed at 550 nm (shown in Fig.3) which shows an excitation maxima at 403nm. Based on the absorption and excitation studies, the excitation wavelength (λ_{exc}) was fixed at 403 nm for the PL recording of the samples. Moreover, due to the overlapping of the absorption spectra of S2 and S4 with that of singly doped ZnS sample, the absorption intensities for the ZnS co-doped Sm^{3+} samples enhance.

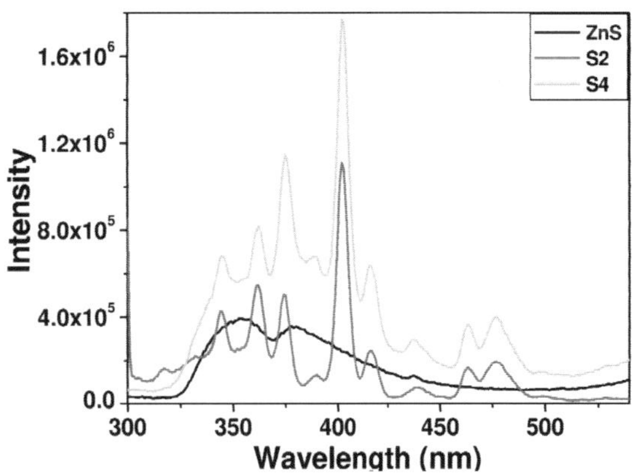

Figure 2: Excitation spectra for singly doped ZnS, S2 and S4 samples for emission wavelength fixed at 596 nm

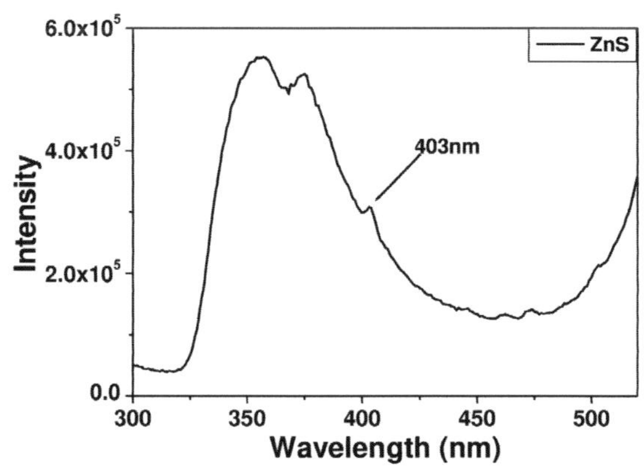

Figure 3: Excitation spectra for singly doped ZnS for emission wavelength fixed at 550nm.

Fig. 4 reveals the PL spectra of the Sm^{3+} samples co-doped with and without ZnS. All the samples show four major PL peaks of Sm^{3+} at 560 nm ($^4G_{5/2}\rightarrow{}^6H_{5/2}$), 596 nm ($^4G_{5/2}\rightarrow{}^6H_{7/2}$), 642 nm ($^4G_{5/2}\rightarrow{}^6H_{9/2}$) and 702 nm ($^4G_{5/2}\rightarrow{}^6H_{11/2}$) [9, 13] respectively. It is observed that the intensities of all the PL peaks increase with the increase in ZnS concentration; the highest enhancement of the intensity being nearly one order of magnitude. The monotonous increase in the PL intensity with the ZnS concentration suggests that further enhancement of PL can be achieved with higher ZnS. Fig. 5 shows the PL spectra for singly doped ZnS sample whereas Fig. 6 shows the emission spectra of S4 in three different λ_{exc} viz. 363 nm, 375 nm and 403 nm. For all the excitations (refer Fig. 6), though the PL peaks appear at the same positions, the intensity is highest for λ_{exc} 403 nm due to its high oscillator strength. The PL intensities, Yields and FWHMs for all the observed PL maxima are shown in table 1.

The mechanism leading to the enhancement of PL with the addition of ZnS is depicted by the energy level diagram shown in Fig. 7. The ZnS NPs possess different energy levels owing to the presence of defect states. The level I_{Zn} originates due to Zn interstitials which is at an energy ~24814 cm^{-1} ($\approx$ 403nm) above the valence band. The level V_{Zn} arises due to Zn vacancy. When the sample is excited at 403nm, two types of absorption take place. In the first mechanism, the Sm^{3+} ions absorb the excitation energy leading to the transition $^6H_{5/2}\rightarrow{}^6P_{3/2}$. The excited electrons at $^6P_{3/2}$ then radiate non-radiatively and populate the $^4G_{5/2}$ level. In the second case, the ZnS NPs absorb the 403 nm excitation

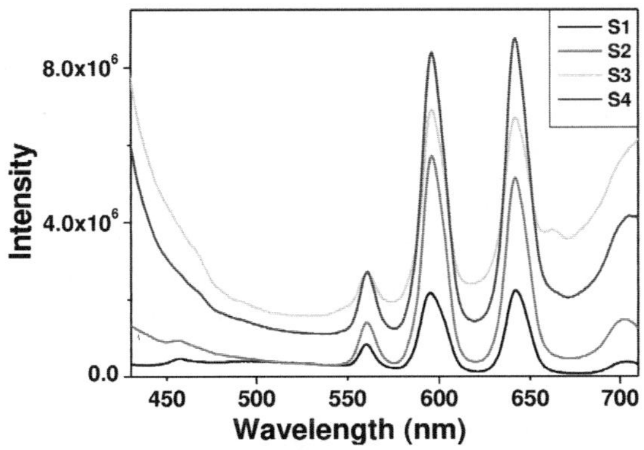

Figure 4: PL spectra for S1, S2, S3 and S4 at 403nm excitation wavelength.

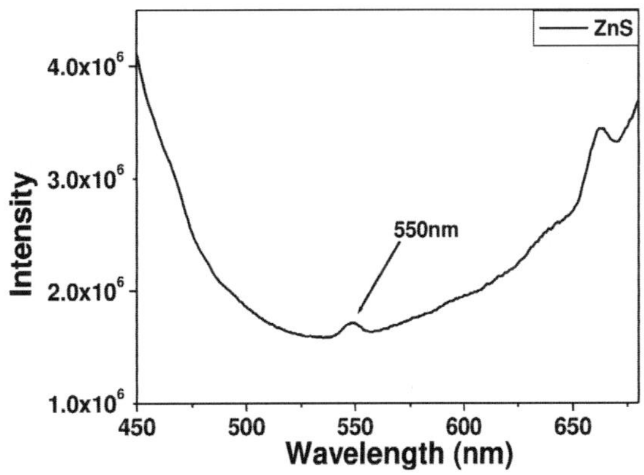

Figure 5: PL spectra of singly doped ZnS at 403nm excitation wavelength.

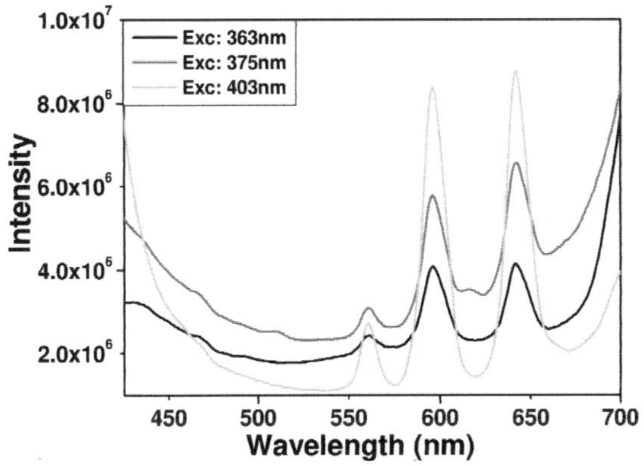

Figure 6: PL spectra of sample S4 for different excitation wavelengths.

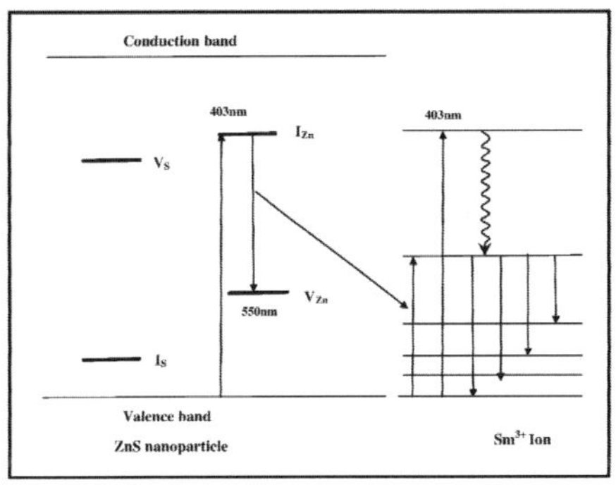

Figure 7: Energy level diagram depicting the PL mechanism.

Table 1: PL intensities, Yield and FWHM for samples S1, S2, S3 and S4 under 403 nm excitations.

PL maxima	PL Intensity			
	S1	S2	S3	S4
560	8.39x10^5	1.4x10^6	2.68x10^6	2.72x10^6
596	2.17x10^6	5.72x10^6	6.91x10^6	8.39x10^6
642	2.24x10^6	5.15x10^6	6.71x10^6	8.76x10^6
	Yield			
560	1.32x10^7	2.32x10^7	6.30x10^8	5.78x10^7
596	4.05x10^7	9.77x10^7	1.66x10^8	1.66x10^8
642	3.877x10^7	9.47x10^7	1.74x10^8	1.98x10^8
	FWHM (nm)			
560	13	13	30	19
596	17	15	20	16
617	15	15	26	17

and lift the valence band electrons to I_{Zn}. The electrons at I_{Zn} recombine with the holes at V_{Zn} thus emitting radiation at 550nm (refer Fig. 5) [16, 17]. This emitted energy is subsequently absorbed by the Sm^{3+} ions in the ground state which further enhances the population of $^4G_{5/2}$. Radiative transition from the $^4G_{5/2}$ to $^6H_{5/2}$, $^6H_{7/2}$, $^6H_{9/2}$, $^6H_{11/2}$ respectively produces the PL maxima at 560nm, 596nm, 642nm and 702nm. The enhancement of PL intensity with concentration of ZnS can be attributed to the enhancement of the population of ZnS NPs around the Sm^{3+} ions as well as the decrease in the distance between the ZnS NPs and Sm^{3+} ions which enhances the efficiency of energy transfer from ZnS NPs to Sm^{3+} ions.

4. Conclusion

In the present work, the absorption and PL behavior of Sm^{3+} in presence of ZnS NPs in methanol medium has been studied. A monotonous enhancement in PL intensities and yield with the addition of ZnS Nps were observed for the studied samples. The enhancement is attributed to the effective increase of Sm^{3+} absorption because of the addition of ZnS NPs. The study shows that the efficiency can be further enhanced with the increase in ZnS concentrations.

References

[1] Binnemans K. 2009 *Chem. Rev.* **109** 4283.
[2] Jayasimhadri M., Cho E., Jang K., Lee H. and Kim S. 2008 *J. Phys. D: Appl. Phys.* **41** 175101.
[3] Ricci P., Carbonaro C, Lehmann A., Congiu F., Puxeddu B., Cappelletti G.and Spadavecchia F. 2013 *J Alloys Compd.* **561** 109.
[4] Jose G., C. Joseph, Ittyachen M. and Unnikrishnan N. 2006 *Opt. Mater.* **29** 1495.
[5] Wageh S., Ahmed A., Ghamdi A. and Yakuphanoglu F. 2013 *J. Sol-Gel Sci Technol.* **66** 443.
[6] Arago J., Lopez B., Cordoncillo E, Escribano P., Pelle F., Viana B. and C. Sanchez 2008 *J. Mater. Chem.* **18** 5193.
[7] Sun L.,Yan C., Liu C., Liao C., Li D. and Yu J. 1998 *J. Alloys Compd.* **275** 234.
[8] Chen L., Zhang J., Lu S., Ren X and Wang X. 2005 *Chem Phys Lett.* **409** 144.
[9] Tohda T., Fujita Y., Matsuoka T. and Abe A 1985 *Appl. Phys. Lett.* **48** 95.
[10] Herrmann A., Kuhn S., Tiegel M., Russel C., Körner J., Klopfel D., Hein J. and Kaluza M. 2014 *J. Mater.Chem.C* **2** 4328.
[11] Dillip G., Kumar P. and Raju B 2013 *J. Lumin.* **134** 333.
[12] Deun R., Binnemans K., Warrand C. and Adam J. 1999 *Proc. SPIE 3622, Rare-Earth-Doped Materials and Devices III*, 175.
[13] Reddy C., Naresh V., Babu B. and S. Buddhudu 2014 *Adv. Mater. Phys. Chem.* **4** 165.
[14] Suhasini T., Kumar J., Sasikala T., Jang K., Lee H., Jayasimhadri M., Jeong J., Yi S. and Moorthy L. 2009 *Opt. Mater.* **3** 1167.
[15] Bokatial L. And Rai S 2012 *J. Opt.* **41** 94.
[16] Taherian M., Alvani A., Shokrgozar M., Salimi R., Moosakhani S., Sameie H., and Tabatabaee F. 2014 *Electron. Mater. Lett.* **10** 393.
[17] Wang X., Shi J., Feng Z., Li M. and Li C. *Phys. Chem. Chem. Phys.* **13** 4715.

Morphology of Colloidal Particles Dispersed in Nematic Solvent

Anupam Kumar*, Biplab Kumar Mandal and Pankaj Mishra

Condensed Matter Theory Laboratory, Department of Applied Physics,
Indian Institute of Technology (ISM) Dhanbad, India

E-mail: *anupamkumar@ap.ism.ac.in

Abstract. We have studied a system of spherical colloidal particles suspended in nematic liquid crystal confined to a two-dimensional plane. The dispersed colloidal particles pervert the uniform orientation of nematic resulting in topological defects. This small change in director field induces elastic interaction in the system. Considering the system exhibiting octopolar symmetry, the interaction of the particles can be described by octopole-octopole interaction potential which on some suitable scaling has the form, $\beta u(r) \approx \Gamma/r^7$, where Γ is dimensionless interaction strength parameter. We have calculated the pair correlation function and radial distribution function of the system by employing Roger-Young's integral equation theory, where the mixing parameter α, is obtained by demanding the consistency in pressure via virial and compressibility routs. With the increase in interaction strength, the system is found to become more ordered.

1. Introduction

Liquid crystals were discovered in 1888 by an Austrian botanist Friedrich Reinitzer [6] while studying the melting behaviour of cholesteryl benzoate. He noticed that the turbid fluid formed at 145.5 ^{0}C transformed to a conventional clear liquid at 178 ^{0}C. He concluded that a new state of matter had been discovered that was intermediate between the crystalline solid and isotropic liquid phases. This phase was later termed liquid crystal by Otto Lehmann [1] who performed the first polarized optical microscopic measurements on liquid crystals. The materials which exhibit this phase are called mesogens [2, 3]. The dispersion of colloidal particles in nematic liquid crystal, also known as nematic colloids has evoked potential interest among the researchers over last few decades because of its wide applications mainly as a biological sensors [4-7], photonic Crystal devices [8], topological memory devices [9,10] etc. This inclusion of particles perverts the long range orientational order of nematic liquid crystal which is marked by the presence of director deformation mediated elastic interaction in the system. Such contortion in director field can be described by Laplace equation leading to the multipole expansion of the director field which is quite similar to electrostatic multipole expansion [11]. These interactions are found to depend upon the nematic director profile and accordingly system can be characterized by dipole-dipole and quadrupole-quadrupole interaction potential [12-14].

However, Chernyshuk et al. has shown that octopole-octopole interaction potential comes into picture when the system carry hedgehog director configuration [15]. Apparently, it has been experimentally [16] observed that the intermolecular interaction in a system of charged colloidal particles attached to water-nonpolar fluid (such as oil) interface in the presence of an external electric field E_o directed normal to the interface, is governed by the octopolar moments leading to a weak short range repulsion. The pair correlation function contains the structural information of fluid. In this paper, our main aim is to calculate the pair correlation function of the system by considering the octopole term as a leading contributor to the interaction potential. We have used Roger-Young's integral equation theory to calculate the pair correlation functions at various grid points of coupling strength parameter. We found notable changes in pair correlation peak with the variation of the characterizing parameters of the system. The paper is organized as follows : in Section 2, we have describe the interaction potential and model of the system followed by the details of the liquid integral equation theory along with the solution procedure in Section 3 and the result in section 4. We finally conclude in Section 5.

2. Model Potential

Let us consider a system of N axially symmetrical colloidal particle dispersed in a nematic liquid crystal confined to a two-dimensional plane of area A. The particle are assumed to induce deformations of the director in the perpendicular directions (homeotropic anchoring) and makes the director field from the undeformed n = (0, 0, 1) (n||z) to deformed n = (n_x, n_y, 1) one. The symmetry of the director deformation mediated elastic interaction is assumed to be such that the particles interact via octopole-octopole interaction potential given by the expression [15-17]

$$\beta U = \beta K \frac{\Omega^2}{r^7} \tag{1}$$

Where $\beta = 1/k_B T$, with k_B being the Boltzmann constant and T is the absolute temperature. Ω is the strength of the elastic octopole moment of the particle, K is Frank elastic constant of the nematic and r is the separation between the colloidal particles. By introducing the average interparticle separation $a = 1/\sqrt{\rho}$, where ρ is number density, as characteristic length scale of the system, we can define the dimensionless interaction strength parameters

$$\Gamma = \beta K \Omega^2 \rho^{7/2} \tag{2}$$

using Eq. (2), we can write Eq. (1) as

$$\beta U = \frac{\Gamma}{x^7}; \qquad \text{With } x = r/a\ (= r^*) \tag{3}$$

interaction strength Γ can be expressed in terms of the dimensionless scaled density ρ^* and temperature t^* as

$$\Gamma = \frac{(\rho^*)^{7/2}}{t^*} \tag{4}$$

where $\rho^* = l^2\rho$ and $t^* = k_B T/\varepsilon$. Here, l $(=\Omega^2/a)$ is the natural length scale and ε (=Ka) is the natural energy scale.

3. Integral equation theory

The determination of the pair structure of a uniform fluid amounts to the calculation of the radial distribution function (RDF) g(r) and the direct correlation function (DCF) c(r). By introducing the total correlation function h(r) = g(r) – 1, the former two quantities are connected via the Ornstein Zernike (OZ) relation

$$h(r) = c(r) + \rho \int d^D r' c(|r - r'|) h(r') \tag{5}$$

where D is the dimensionality of the space which is 2 in the present case. The OZ equation is exact and couples two unknown functions, h(r) and c(r). In order to determine them, another equation connecting these two unknowns, is needed. The total correlation function can be expressed in terms of the direct correlation function as

$$h(r) = exp[\beta u(r) + h(r) - c(r) + B(r)] - 1 \tag{6}$$

where u(r) denotes the pair potential and B(r) is the so-called bridge function.[18] By introducing certain approximations for the bridge function, an approximate equation is obtained which is known as a closure. It is the closure through which interparticle interaction enters into the formulation. Both the RDF as well as the DCF can be obtained by solving Eqs. (5) and (6) self-consistently. By setting the bridge function equal to zero, we obtain the hypernetted chain (HNC) closure, which has proved to yield good results for long-ranged pair potentials in three dimensions. The Percus-Yevick (PY) closure is obtained by choosing the following expression for the bridge function:

$$B_{PY}(r) = -[h(r) + 1 - c(r)] + 1 + ln[h(r) + 1 - c(r)] \tag{7}$$

The PY closure furnishes accurate results for short-range interactions (such as the hard-sphere potential) in three dimensions. A particularly reliable closure relation is provided by the Rogers Young [19] that 'interpolates' between the two and has the form

$$h(r) = exp[-\beta u(r)]\left[1 + \frac{exp[f(r)\{h(r)-c(r)\}]-1}{f(r)}\right] - 1 \tag{8}$$

where f(r) [= 1− exp(-αr)] is called mixing functions which include a tunable parameter α that is chosen in order to enforce thermodynamic consistency by guaranteeing the equality between the virial and fluctuational compressibility. In this work we have employed the Rogers-Young closure to solve the OZ equation. To solve OZ equation, we first transform it into its Fourier form by using Fourier-Bessel transform of a function p(r) in 2D [20] defined as

$$p(k) = 2\pi \int_0^\infty p(r)J_n(kr)rdr \tag{9}$$

$$p(r) = \frac{2\pi}{(2\pi)^2}\int_0^\infty p(k)J_n(kr)kdk \tag{10}$$

where $J_n(kr)$ are the Bessel functions of first kind. Putting Fourier transforms of h and c in Eq. (5) we get

$$h'(k) = c'(k) + \rho c'(k) h'(k) \tag{11}$$

we then solve Eq. (11) and Eq. (8) iteratively at several grid points of interaction strength Γ. The following steps are required to solve these equations numerically:

1. Firstly we take initial guess for $c(r)^i$.
2. Then by using Eq. (9), we transform it into k-space $c'(k)^i$.
3. On substituting the value of $c'(k)^i$ in Eq. (11), we obtain $y'(k)^i$ (= $h'(k)^i - c'(k)^i$).
4. Putting the value of y(r) (after transforming $y'(k)$ to y(r) by using Eq. (10)) in Eq. (8), we get $c(r)^{new}$.
5. We repeat steps (2-4) until convergence is obtained i.e.,
$$\left|c(r)^{new} - c(r)^i\right|_{max} \leq 10^{-3} \tag{12}$$
6. If we obtain the convergence then we take it as a final solution otherwise we take a new input $(i+1)^{th}$ by mixing the ith and the 'new' values such as
$$c(r)^{i+1} = \alpha c(r)^i + (1-\alpha)\, c(r)^{new} \tag{13}$$

and steps (2-5) are repeated.

Here, the value of α varies between 0 and 1. Now if we obtain the value of h(r) then we can get the structure factor S(k) which can be obtained by scattering experiment [21], by using the relation given below

$$S(k) = 1 + \rho h'(k) \tag{14}$$

4. Results and Discussion

In this section we present some representative plots of pair correlation functions directly related to the structural arrangement of particles in the fluid. In Fig. 1, we have plotted the radial distribution function (RDF) at different value of the interaction strength i.e. $\Gamma = 5, 10, 15, 20$. We observe that the radial distribution function (RDF) exhibit structures which become more prominent with the increase in interaction strength. The principle peak represents the location of most populated cell. In other words it indicates the most probable position of the colloidal particles. The probability of occupying the successive cells is seen to decrease with the increase in distance which signals the absence of long range order. We also observe a solder in RDF located at a distance approximately square root times of the distance of first peak. The solder if found to become more prominent with the increase in interaction strength.

Increase in Γ means decrease in temperature which leads to a decrease in particles kinetic energy which in turn results in an enhanced attraction between the particles causing enhanced ordering at shorter distances. For increasing interaction strength, the height of the main peak increases steadily, whereas the peak position does not change. The result is physically reasonable: we can assume that Γ is increased by making some external changes, such as reducing the temperature field or enhancing the strength of some external electric or magnetic field, without changing the total number of particle in the system. Then, the typical interparticle distances within the first coordination shell remain almost unchanged and only the degree of local ordering is affected, due to the increasingly strong repulsions.

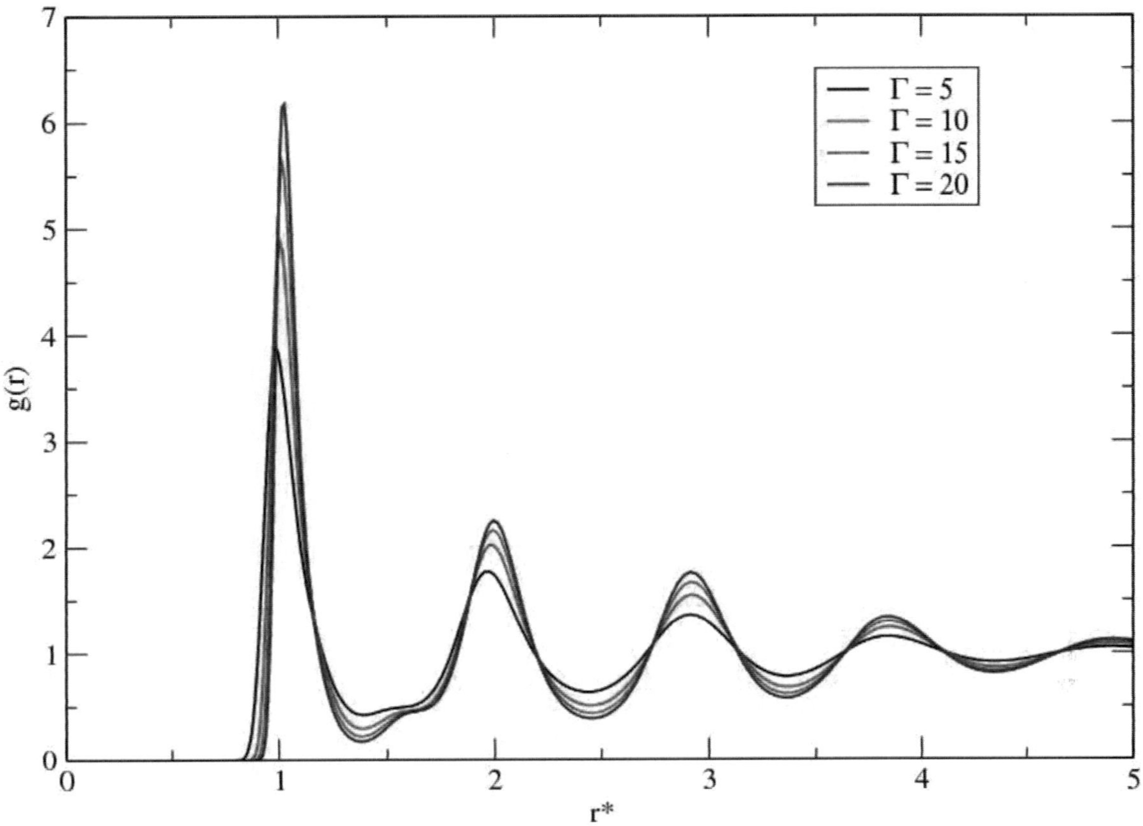

Figure 1. Graph showing variation of g(r) with Γ (interaction strength).

These features of the structure of the system can also be seen in the curves of structure factor shown in Fig.2. Structure factor is an important experimentally measurable physical quantity which is directly related to the RDF through its Fourier transform. The peaks of the structure factor S(k) can be seen to shift slightly to the lower k-side, which confirms the above conclusion that the interparticle interaction becomes more repulsive with the increase in Γ. The peak height of the structure factor curve is also found to increase with the increasing interaction strength. These results are similar to the experimental and computer simulation results of Hoffmann et al [22,23] on a system of superparamagnetic colloidal particles confined to a two dimensional air-water interface.

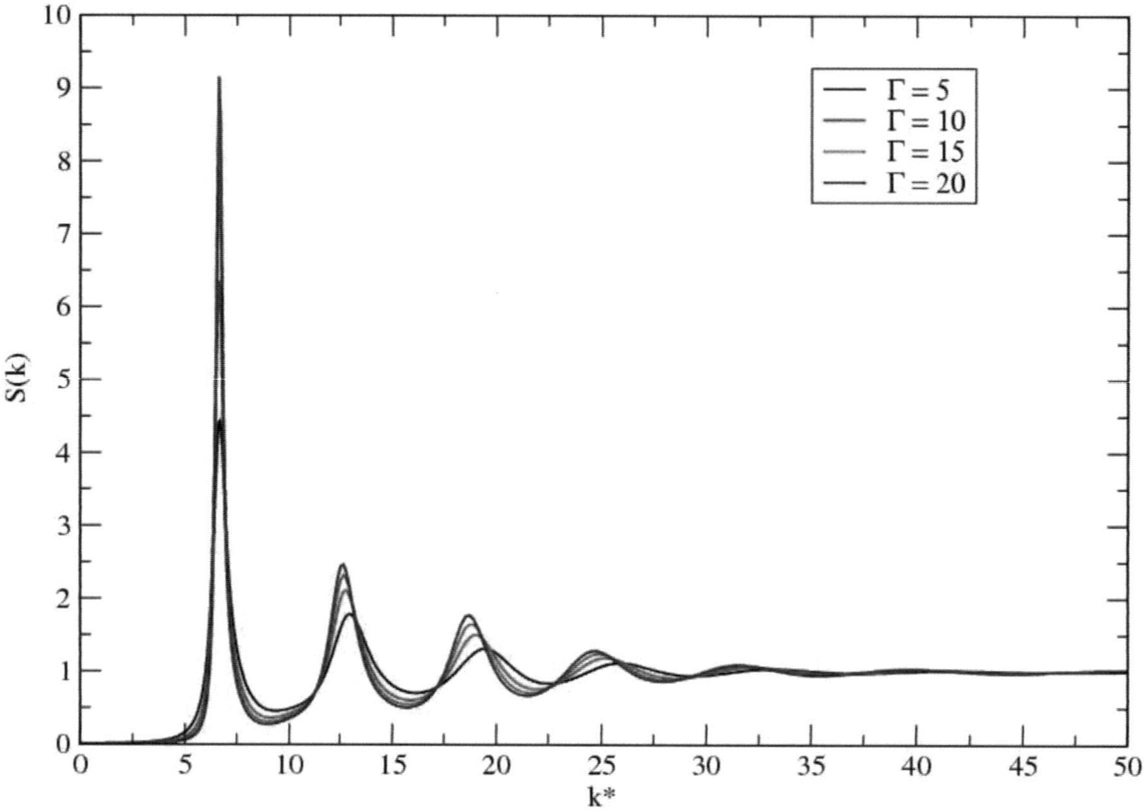

Figure 2. Graph showing variation of S(k) with Γ (interaction strength).

5. Conclusions

We have studied a one component system of nematic colloids confined in a two dimensional plane in which interaction between the particles is governed by repulsive octopole-octopole interaction potential. We have introduced a dimensionless interaction strength parameter to calculate the pair correlation function of the system by using Roger-Young integral equation theory. We observe a significant change in the local ordering of the particle in the system with the increase in the interaction strength parameter. The radial distribution and structure factor curves are found to have a strong short range structure which is seen to enhance both in size and range with the increase in Γ. The pair correlation function calculated in this study will be used in density functional theory of freezing to study the phase diagram of octopoler nematic colloids in two dimensions.

Acknowledgements

Anupam Kumar and Biplab Kumar Mandal would like to thank Indian Institute of Technology (ISM), Dhanbad for providing financial support through a research fellowship.

References

[1] Lehman O 1890 *Z. Phys. Chem.* **5** 427.
[2] Fergason J L 1985 *SID Digest.* **16** 68.
[3] Doane J W, Vaz N A, Wu B G, and Žumer S 1986 *Applied Physics Letters* **48** 269.

[4] Shiyanovskii S V, Schneider T, Smalyukh I J, Ishikawa T, Niehaus G D, Doane K J, Woolverton C J and Lavrentovich O D 2005 *Phys. Rev. E* **71** 020702.
[5] Helfinstine S, Lavrentovich O, and Woolverton C *Lett. Appl. Microbial.* **43** 27.
[6] Hussain A, Pina A and Roque A 2009 *Biosens. Bioelectron.* **25** 1.
[7] Wollman S, Jay G and Orawford G 2007 *Nat. Mater.* **6** 929.
[8] Muševič I 2013 *Phil Trans RSoc A* **371** 20120266.
[9] Porenta T, Cˇopar S, Ackerman P J, Pandey M B, Varney M C M, Smalyukh I I and Žumer S 2014 *Sci. Rep.* **4** 7337.
[10] Lubensky T C, Pettey D, Currier N, and Stark H 1998 *Phys. Rev. E* **57** 610.
[11] Terentjev E M 1995 *Phys. Rev. E* **51** 1330.
[12] Sˇkarabot M, Ravnik M, Žumer S, Tkalec U, Poberaj I, Babiˇc D, Osterman N, and Muˇseviˇc I 2008 *Phys. Rev. E* **77** 031705.
[13] Tovkach O M, Chernyshuk S B and Lev B I 2012 *Phys. Rev. E* **86** 061703.
[14] Noel C M, Bossis G, Chaze A M, Giulieri F and Lacis S 2006 *Phys. Rev. Lett.* **96** 217801.
[15] Chernyshuk S B, Tovkach O M and Lev B I 2014 *Phys. Rev. E* **89(3)** 032505.
[16] Danov K D and Kralchevsky P A 2013 *Journal of colloid and interface science* **405** 278.
[17] Chernyshuk S B *Eur. Phys. J. E* **37** 6.
[18] Hansen J P and McDonald I 2005 *Theory of Simple liquids* 3rd Edition, Academic Press, London.
[19] Rogers F J and Young D A 1984 *Phys. Rev. A* **30** 999.
[20] Talman J D 1978 *J. Comput. Phys.* **29** 35.
[21] Pasquarello A, Sarnthein J and Car R 1998 *Phys. Rev. B* **57** 14133.
[22] Hoffmann N, Ebert F, likos C N, Löwen H and Maret G 2006 *Phys. Rev. Lett.* **97** 078301.
[23] Hoffmann N, likos C N and Löwen H 2006 *J. Phys.: Condens. Matter* **18** 10193.

Photoluminescence and photocatalytic activities of Ag/ZnO metal-semiconductor heterostructure

Bikash Sarma, Sujit Kumar Deb and Bimal K. Sarma*
Nanophysics Laboratory, Department of Physics, Gauhati University, Guwahati-781014, India
*Corresponding author E–mail: sarmabimal@gmail.com

Abstract. Present article focuses on the photocatalytic activities of ZnO nanorods and Ag/ZnO heterostructure deposited on polyethylene terephthalate (PET) substrate. ZnO nanorods are synthesized by thermal decomposition technique and Ag nanoparticles deposition is done by photo-deposition technique using UV light. X-ray diffraction studies reveal that the ZnO nanorods are of hexagonal wurtzite structure. Further, as-prepared samples are characterized by Scanning Electron Microscopy (SEM), Photoluminescence (PL) spectroscopy and UV-Vis spectroscopy. The surface plasmon resonance response of Ag/ZnO is found at 420 nm. The photocatalytic activities of the samples are evaluated by photocatalytic decolorization of methyl orange (MO) dye with UV irradiation. The degradation rate of MO increases with increase in irradiation time. The degradation of MO follows the first order kinetics. The photocatalytic activity of Ag/ZnO heterostructure is found to be more than that of ZnO nanorods. The PL intensity of ZnO nanorods is stronger than that of the Ag/ZnO heterostructure. The strong PL intensity indicates high recombination rate of photoinduced charge carriers which lowers the photocatalytic activity of ZnO nanorods. The charge carrier recombination is effectively suppressed by introducing Ag nanoparticles on the surface of the ZnO nanorods. This study demonstrates a strong relationship between PL intensity and photocatalytic activity.

Keywords: ZnO nanorods, Ag/ZnO heterostructure, Photo-deposition, Exciton-plasmon system, Photoluminescence, Photocatalysis.

1. Introduction

ZnO is a promising wide band gap semiconductor with direct band gap of 3.37 eV at room temperature and has a large exciton binding energy of 60 meV [1]. ZnO nanostructures have lots of applications in various fields such as varistors, electroluminescent displays, luminescence materials and sensors [2]. ZnO has a lot of importance in the field of photocatalysis. Tan *et al* reported that the incorporation of Ag on the surface of ZnO could improve the photocatalytic efficiency of ZnO upto a great extent [3]. Due to the large work function, the Ag nanoparticles can slow down the recombination of photogenerated electron-hole pairs. In addition, the surface plasmon resonance (SPR) of Ag nanoparticles can enhance the solar light absorption efficiency of the Ag/ZnO heterostructure.

Photoluminescence spectroscopy (PL) is a powerful technique to investigate the photophysical properties of semiconductor materials. Moreover, PL spectrum provides information about surface defects, oxygen vacancies, carrier capture efficiency etc. A higher PL intensity indicates higher recombination rate of photo-induced electron-hole pairs and vice-versa. Pronin *et al* reported strong relationship between PL intensity and photocatalytic efficiency [4]. Usually, photocatalyst with lower PL intensity exhibits higher photocatalytic efficiency. This is attributed to the lower recombination rate of photo-induced electron-hole pairs. The semiconductor photocatalysis is environment friendly and cost effective process. The main focus of this work is to study the photophysical and photochemical properties of ZnO nanorods and Ag/ZnO heterostructure deposited on flexible polyethylene terephthalate (PET) substrate. An attempt has been made to increase the visible light absorption of ZnO through the incorporation of Ag nanoparticles in the Ag/ZnO heterostructure. This process might increase the photocatalytic activity of the Ag/ZnO metal-semiconductor system which is strongly related to the luminescence of the exciton-plasmon system.

2. Experimental section

2.1. Synthesis of ZnO nanorods

The ZnO nanorods are prepared on flexible PET substrate using thermal decomposition method. Ultrasonically cleaned PET substrates are placed inside a beaker to which a mixture of 50 ml each of 10 mM zinc-nitrate hexahydrate [$Zn(NO_3)_2 \cdot 6H_2O$] and 10 mM hexamethylene tetramine [$C_6H_{12}N_4$] from Merck are poured. The beaker is placed inside a water bath with constant heating at 90 °C for duration of 3 hours. The prepared film is then rinsed with distilled water and kept inside an oven at 90 °C to dry it up.

2.2. Synthesis of Ag/ZnO heterostructure

Ag/ZnO heterostructures are prepared using photo-deposition technique. The previously grown ZnO is kept in a Petri dish. A mixture of 6 ml ethylene glycol [$HOCH_2CH_2OH$] from Merck, 0.10 ml of 20 mM silver nitrate [$AgNO_3$] from Sigma-Aldrich, and 3 ml distilled water is poured over the ZnO nanorods. The Petri dish is kept under UV light irradiation for 1.5 hours. Four UV lamps of 11 W each (Philips) are used for this purpose. The prepared film is then rinsed with distilled water and dried in an oven at 90 °C.

2.3. Photocatalysis experiment

Methyl orange (MO) solution is used for the photocatalytic degradation using ZnO nanorods and Ag/ZnO heterostructure as catalysts. Required amount of MO powder from Merck is dissolved in distilled water to prepare 50 μM concentration of MO solution. 5 ml of MO solution is added to two beakers one containing ZnO nanorods and the other containing Ag/ZnO heterostructure. The beakers are kept under UV-Vis irradiation. This process is carried out for different irradiation time. The absorbance of the degraded solutions is taken for analysis. The degradation efficiency, η is measured using the relation:

$$\eta = \frac{A_0 - A}{A_0} \times 100\% \tag{1}$$

(1)

where A_0 is the absorbance of the MO solution before illumination and A is the absorbance of the MO solution after irradiation time t.

2.4. Characterization

X-ray diffraction (XRD) data are recorded using an X-PERT PRO X-ray diffractometer with a Cu K_α (1.5406 Å) radiation source at a tube voltage of 40 kV and a current of 35 mA. UV-Vis absorbance spectra are measured using UV-Vis spectrophotometer (CARY EL02086568). SEM images are taken using ΣIGMA-Zeiss Field Emission Scanning Electron Microscope (FE-SEM). The room temperature PL spectra are recorded using Eddinburg FSP920 spectrometer with xenon lamp as an excitation source. The excitation wavelength is set at 300 nm.

3. Results and discussion

3.1. X-ray diffraction studies

XRD patterns of ZnO nanorods and Ag/ZnO heterostructure deposited on PET substrate are shown in Figure 1. The diffraction peaks are indexed to hexagonal wurtzite ZnO (ICDD 36-1451). The Ag diffraction peak is indexed to face centred cubic (fcc) metallic Ag (ICDD 04-0783). XRD patterns

Journal of Physics: Conference Series **765** (2016) 012023 doi:10.1088/1742-6596/765/1/012023

show that both the samples are crystalline in nature. No change of diffraction peaks is observed for ZnO after Ag deposition. The diffraction peaks for ZnO nanorods are found at 2θ values 31.66°, 34.31°, 36.12°, 47.50°, 56.46°, 62.84°, 66.34°, 67.88°, 69.10°, and 77.05° corresponding to the diffraction planes (100), (002), (101), (102), (110), (103), (200), (112), (201), and (201) respectively. The intensity of (100) peak is maximum which indicates most of the ZnO nanorods are oriented along *a*-axis. The intensity of Ag diffraction peak is very weak and it is seen at 2θ value 38.03° and the corresponding diffraction plane is (111).

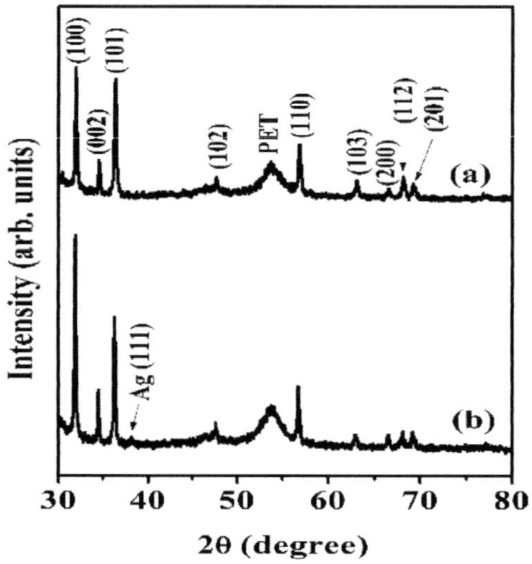

Figure 1. XRD patterns of (a) ZnO nanorods, (b) Ag/ZnO heterostructure.

The crystallite size of Ag nanoparticles is calculated using the Scherrer formula [6]

$$t = \frac{k\lambda}{\beta \cos\theta} \tag{2}$$

Here, t is the crystallite size, k is the Scherrer constant, λ is the wavelength of X-ray, β is the full width of the peak at half-maximum and θ is the half of the diffraction angle. The crystallite size of the Ag nanoparticle is found to be 18 nm.

3.2. FE-SEM studies

Figure 2. FE-SEM images of (a) ZnO nanorods, (b) flower-like structure formed by ZnO nanorods, (c) Ag/ZnO heterostructure, and (d) higher magnification image of Ag/ZnO heterostructure.

In order to investigate the morphologies of the as prepared ZnO nanorods and Ag/ZnO heterostructure FE-SEM investigation is carried out. Figure 2(a) shows the FE-SEM image of the ZnO nanorods. The rods are oriented along different directions. The average length and diameter of the ZnO nanorods with well defined hexagonal facets is found to be 6 μm and the 700 nm respectively. The rod has tapered ends with diameter 500 nm. The flower like structure formed by the ZnO nanorods is shown in Figure 2(b). Figure 2(c) shows the SEM image of the Ag/ZnO heterostructure. Ag nanoparticles are seen on the surface of the nanorods. Self nucleated or isolated Ag nanoparticles are hardly seen. Ag nanoparticles are almost entirely covering the ZnO nanorods. Free areas on the ZnO nanorods are very less. A higher magnification image of Ag/ZnO heterostructure is shown in Figure 2(d). The average sizes of the Ag nanoparticles are 60 nm. Slight variations in the sizes of the Ag nanoparticles are observed.

3.3. UV-Vis spectroscopy studies

The absorption properties of ZnO nanorods and Ag/ZnO heterostructure are investigated with UV-vis spectroscopy in the wavelength range 200-800 nm. Figure 3 shows the absorption spectra for both ZnO nanorods and Ag/ZnO heterostructure. Both the samples show strong absorption in the UV range. A broad band centering at 420 nm is seen in the absorption spectra of Ag/ZnO heterostructure. This is the characteristic of surface plasmon resonance absorption of Ag nanoparticles. The optical band gap of ZnO nanorods is found to be 3.20 eV which is smaller than the band gap of bulk ZnO. The band gap narrowing of ZnO nanorods is attributed to the formation of oxygen vacancies [7].

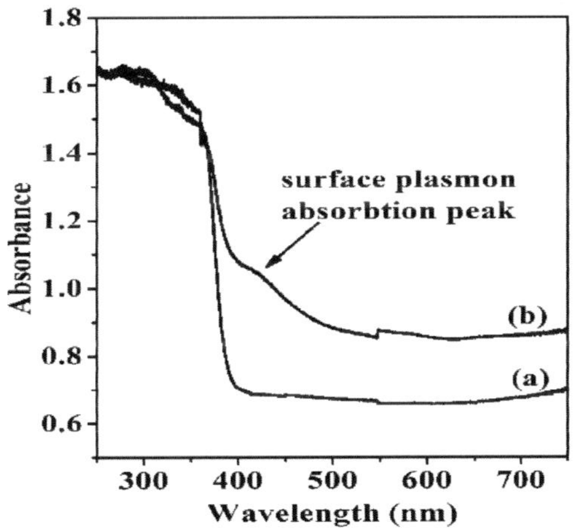

Figure 3. UV-Vis absorbance spectra of (a) ZnO nanorods, (b) Ag/ZnO hetrostructure.

3.4. Photoluminescence studies

Figure 4 shows the PL spectra of ZnO nanorods and Ag/ZnO heterostructure at an excitation wavelength of 300 nm. Figure 4(a) shows two emission peak centered at 391 nm and 447 nm. The former is attributed to the near band edge emission (NBE) and the latter is defined as the deep level emission (DLE) [8,9]. The NBE originates due to the recombination of free excitons formed when excited by radiation of energy greater or equal to the band gap energy. The DLE is a result of several intrinsic defects such as zinc interstitials (Zn_i), interstitial oxygen (O_i), zinc vacancies (V_{Zn}), oxygen vacancies (V_O), etc [10,11]. The DLE is mainly attributed to electron transition from shallow donor level formed by Zn_i to the top of valance band (V_{Zn}) [12]. Several defects are responsible for the high intense long range green emission from 500-700 nm.

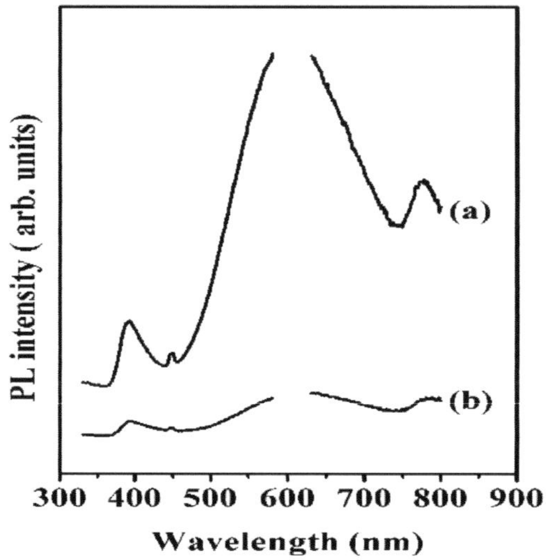

Figure 4. Photoluminescence spectra of (a) ZnO nanorods, (b) Ag/ZnO heterostructure.

The combination of O_i and V_{Zn} emissions results in the red band forming at 773 nm [13]. The intensity of NBE, DLE and the green emission decreases for Ag/ZnO heterostructure (Figure 4(b)) with no change in peak position. This accounts for the interaction between excitons of ZnO and surface plasmons of Ag nanoparticles. The recombination rate of electrons and holes decreases which results in low PL intensity [14]. The Ag nanoparticles attached over ZnO nanorods act as traps for photogenerated electrons and quench the PL emission. As the ZnO nanorods and Ag nanoparticles are in direct contact, electron transfer takes place between the energy levels of ZnO and Ag.

3.5. Photocatalytic activity studies

The photocatalytic performance of Ag/ZnO heterostructure is superior to ZnO nanorods. Figure 5 shows the absorbance spectra of MO solution using ZnO nanorods and Ag/ZnO heterostructure as catalyst for 8 hours of irradiation time. The photodegradation efficiency is 37.17% for ZnO as catalyst and 40.46% for Ag/ZnO as catalyst for 8 hours UV-Vis irradiation.

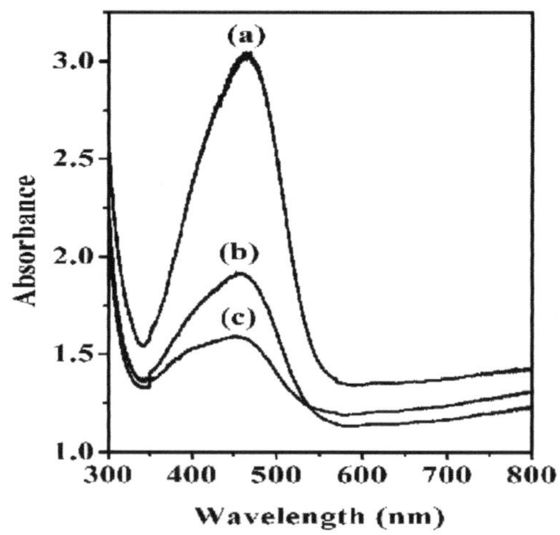

Figure 5. UV-Vis absorbance spectra of MO solution with (a) no catalyst, (b) ZnO nanorods as catalyst, and (c) Ag/ZnO heterostructure as catalyst.

The degradation of MO follows first order kinetics. Therefore the rate equation is:

$$\ln\left(\frac{C}{C_0}\right) = k't \tag{3}$$

Here k' is the first order rate constant, C_0 is the initial concentration of MO solution, C is the concentration of the MO solution after irradiation and t is the irradiation time. The (C/C_0) versus t curves for MO solution with ZnO nanorods and Ag/ZnO heterostructure are shown in Figure 6. The degradation rate constants for ZnO nanorods and Ag/ZnO heterostructure are found to be 0.0007 min^{-1} and 0.0009 min^{-1} respectively.

Electrons and holes are produced as a result of UV irradiation in ZnO. The photogenerated electrons react with dissolved oxygen to form superoxide anion radicals ($^{\bullet}O_2^-$) and holes react with hydroxide ions (OH^-) to produce hydroxyl radicals ($^{\bullet}OH$) [9]. The presence of Ag nanoparticles acts as sink for electrons in the Conduction Band (CB) of ZnO and decreases their recombination rate with the holes allowing the photocatalytic process to continue further. These reactions result in decolorization of the MO solution. The reactions as proposed by Zheng *et al* [14] are listed below:

$$\begin{aligned} &Ag \rightarrow Ag^+ + e^- \\ &e^- + O_2 \rightarrow {}^{\bullet}O_2^- \\ &ZnO + h\nu \rightarrow e_{cb}^- + h_{vb}^+ \\ &e_{cb}^- + Ag^+ \rightarrow Ag \\ &h_{vb}^+ + OH^- \rightarrow {}^{\bullet}OH \end{aligned} \tag{4}$$

PL studies suggest the decrease in intensity with the incorporation of Ag nanoparticles. Moreover, photocatalytic studies confirm the increase in photocatalytic activity due to the presence of Ag nanoparticles. These two facts describe an inherent relationship between PL and photocatalytic activity. The lower the PL intensity, the greater is the photocatalytic activity and vice-versa.

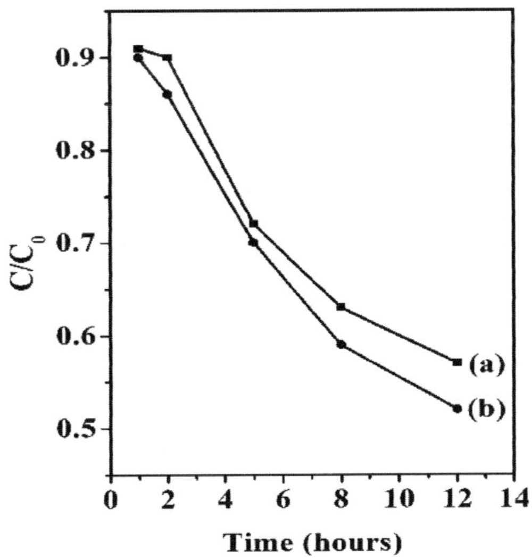

Figure 6. Plot of normalized concentration of MO solution versus irradiation time with (a) ZnO nanorods and (b) Ag/ZnO heterostructure as catalyst.

4. Conclusions

ZnO nanorods are prepared by thermal decomposition method and Ag deposition is done by photo-deposition technique. XRD studies show that the as prepared ZnO nanorods are of hexagonal wurtzite structure. Ag diffraction peak in the XRD pattern of Ag/ZnO heterostructure is very weak and this can be indexed to face centered cubic (fcc) metallic Ag. FE-SEM studies reveal that the ZnO nanorods are randomly aligned and this forms flower-like structures. In the Ag/ZnO heterostructure Ag nanoparticles are seen on the surface of the ZnO nanorods. The band gap narrowing of ZnO nanorods is attributed to the oxygen vacancies. Quenching in the PL emission has been observed in the Ag/ZnO heterostructure. This could be attributed to the exciton plasmon interaction in the Ag/ZnO heterostructure. The photocatalytic activity of the Ag/ZnO heterostructure has been found to be better than that of ZnO nanorods. A strong relationship exists between PL intensity and photocatalytic activity. When PL intensity is weak, the photocatalytic efficiency is high and vice versa. Again the degradation rate of MO increases with increase in irradiation time.

Acknowledgements

Authors acknowledge the Department of Science and Technology (DST) for the financial support under the Grant No. SB/FTP/PS-36/2014. Authors gratefully thank Institute of Advance Study in Science and Technology (IASST), Central Instruments Facility, Indian Institute of Technology Guwahati, and Sophisticated Analytical Instruments Facility, Gauhati University for providing materials characterization facilities.

References

[1] Mclaren A, Solis J V, Li G and Jsang S C 2009 *J. Am. Chem. Soc,* **131** 12540-12541
[2] Arya S K *et al* 2012 *Analytica Chimica Acta,* **737** 1-21
[3] Tan T et al 2008 Materials Chemistry and Physics **111** 305-308
[4] Pronin I.A. *et al* 2013 *Semiconductors* **48** 842-847
[5] Hoffmann M. R., Martin S.T., Choi W. and Bahnemann D.W. 1995 *Chem. Rev.,* **95** 69-96
[6] Chen C., Liu J., Liu P. and Yu B. 2011 *Advances in Chemical Engineering and Science* **1** 9-14
[7] Ansari S.A. *et al* 2013 *Nanoscale* **5** 9238-9246
[8] Hossein S.M., Sarsari I.A., Kameli P. and Salamati H. 2015 *J. Alloys Comp.* **640** 408-415
[9] Karunakaran C., Jayabharathi J., Jayamoorthy K. and Vinayagamoorthy P. 2012 *Journal of Photochemistry and Photobiology A: Chemistry* **247** 16-23
[10] Lin B., Fu Z. and Jia Y. 2012 *Appl. Phys. Lett.* **79** 943-945
[11] Silambarasan M., Saravanan S., and Soga T. 2015 *Int. Journal of Chemtech Research* **7** 1644-1650
[12] Ding J., Yan X., and Xue Q. 2012 *Materials Chemistry and Physics* **133** 405- 409
[13] Roqan I.S. *et al* 2015 *Jour. of Appl. Phys.* **117** 0739041-0739046
[14] Zheng Y. *et al* 2007 *Inorganic Chemistry* **46** 6980−6986

Integral Equation theory of a system of nematic quadrupoles

Biplab kumar Mandal*, Anupam Kumar and Pankaj Mishra

Condensed matter theory lab, Department of Applied Physics
Indian Institute of Technology(ISM), Dhanbad, India

Abstract—We have studied a system of nematic quadrupoles interacting via inverse power potential of the form $U \approx (\sigma/r)^5$ (σ is the diameter of the particles and r is the separation between them), both in 3-dimension and 2-dimension . Pair correlation function has been calculated by solving Rogers–Young (RY) integral equation theory. Thermodynamic consistency has been obtained ensuring the equality of pressure calculated by virial and compressibility routes.

E-mail: biplab.ism@gmail.com

Introduction

Over last 30 years Integral Equation Theories are most sophisticated tool to investigate the topology of complex fluids[1,2].The whole integral equation theory stand on the Ornstein-Zernike (OZ) equation which is a relation between total correlation function h(r) and the direct correlation function c(r). In order to solve OZ, another equation connecting c(r) and h(r) is needed. Whereas OZ is an exact relation, the second one is an approximation and known as closure relation. Many closure relations are available in literature[3,4], among them Percus-Yevick (PY) and hyper-netted Chain (HNC) closures are the most frequently used relations. However Monte Carlo (MC) and Molecular Dynamics (MD) computer simulations studies have revealed various shortcomings of these equations and hence efforts have been made to develop more accurate closure relations [5].

The main problem in PY and HNC is that both are effected by thermodynamic inconsistency that means getting different results of the same quantity from different thermodynamic path. In order to overcome this problem of inconsistency, Rogers and Young proposed a new closure relation between c(r) and h(r) with an adjustable parameter to help in interpolating the PY and HNC values and to insure the consistency of the pressure calculated by various routs[7]. The results obtained by using Rogers and Young (RY) approximations are far accurate than the above mentioned approximations and also compares well with the results of MC and MD simulations.

In this paper we have studied a worked on a simple one component system which is interacting a potential which varies as the 5th power of the inverse interparticle separation. We have employed Roger-Young closure and have calculated pair structures of the fluid phase as a function of various strength parameters of the potential. The paper is arranged as follows: In section II, we discuss the potential model. This will be followed by a brief discussion of the details of the integral equation theory and its solution procedure in section III. In section IV we will present and discuss our results. Finally, we conclude in section V.

II. Model Potential

We consider a system consisting of *N* spherical colloidal particle suspended in a nematic solvent. The interaction of the liquid crystal molecules do the surface of colloids gives rise to a distortion in nematic director field. The colloidal particles are considered to interact with each other via the director distortion mediated elastic interaction which varies as $U(r) = \epsilon\,(\sigma/r)^n$. In the condition of weak homeotropic surface anchoring the value of n can be taken to be equal to 5. It corresponds to the case of interaction between two elastic quadrapoles in analogy with the interaction between the electric quadrupoles.

We define dimensionless interparticle separation by $r^* = r/a$ where $a = (3/4\pi\rho)^{(\frac{1}{3})}$ is the ion-sphere radius. Putting the value of $\rho_0 = \frac{\sqrt{2}}{\sigma_0^3}$ (number density) and $\rho = N/V$ and $r = r/a$ the potential can be expressed in the form

$$\beta\, u(r) = \Gamma / r^5 \qquad (1)$$

Where $\Gamma = (4\pi\sqrt{2}z/3)^{5/3}$ with $z = (N\sigma^3/\sqrt{2}v)(\varepsilon/K_BT)^{\frac{3}{5}}$, is the coupling parameter determining the strength of interaction.

III. Integral equation theory

Integral equation theory involves in writing an exact relation between the total pair-correlation function h(r) and direct pair-correlation function c(r). Such a relation is called OZ, which is written as [6]

$$h(r) - c(r) = \rho \int h(r')c(r - r')dr \qquad (2)$$

Where h(r) is related to the radial distribution function g(r) as h(r) =g(r)-1. In practice OZ equation is solved by writing it into k-space by using the Hankel transforms of the pair-correlation functions. The general form of Hankel transform of a function f(r) written as,

$$f(k) = 4\pi(-p)^m \int_0^\infty dr\, r^2 f(r) J_m(kr) \qquad (3)$$

$$f(r) = \frac{4\pi}{(2\pi)^3}(-p)^m \int_0^\infty dk\, k^2 f(k) J_m(kr) \qquad (4)$$

where $J_m(kr)$ is the spherical Bessel function of order m.

We have solved OZ, by using Roger- Young closure written as,

$$h(k) = c(k) + \rho\, h(k)\, c(k) \qquad (5)$$

We have solved OZ, by using Roger- Young closure written as,

$$g(r) = e^{-\beta r}\left[1 + \frac{e^{f(r)\gamma(r)} - 1}{f(r)}\right] \qquad (6)$$

Where $\gamma(r) = h(r) - c(r)$ and f(r) is the mixing function has a limit $0 \le f(r) \le 1$.

It is written as, $f(r) = 1 - \exp^{-\alpha r}$, where α is the mixing parameter.

In order to determine the value of mixing parameter, we start by choosing its value arbitrarily and solve the integral equation theory. The resulting pair-correlation function are used to calculate the pressure from virial route using the relation

$$p = \rho\, KT\left[1 - \frac{\rho}{2dKT}\int \frac{du}{dr} r g(r) dr\right] \qquad (7)$$

where d is the dimension of the space and the excess pressure from compressibility route can be written as,

$$\beta \frac{dP}{d\rho} = 1 - 4\pi\rho \int c(r) r^2 dr \qquad \text{for 3D} \tag{8}$$

$$\beta \frac{dp}{d\rho} = 1 - 2\pi\rho \int c(r) r^2 dr \qquad \text{for 2D} \tag{9}$$

In order to obtain the thermodynamic consistency, pressure calculated from these two routes must be equal.

Algorithm

The numerical solution of OZ and RY-closure is performed following the iteration cycle,

$c(r)^j \rightarrow c(r)^j \rightarrow (\text{o-z})\gamma(k)^j \rightarrow \gamma(r)^j \rightarrow (\text{closure})\ c(r)^{new}$

where superscript j denote the jth iterate. Iterative procedure involves the following steps:

a. First we make an initial guess for $c(r)^j$.
b. This is transformed to k space by $c(k)^j$ Hankel transform [Eq.3].
c. This $c(k)^j$ is then substituted in the OZ equation[Eq.2] to get the value of $\gamma(k)^j$ $(=h(k)^j - c(k)^j)$
d. By using inverse Hankel transform [Eq.4] step $\gamma(k)^j$ converted to $\gamma(r)^j$.
e. This $\gamma(r)^j$ is substitute in the RY closure [Eq.6] we get the new $c(r)^{new}$.
f. The above steps (b) to (e) are repeated until a fixed point is reached satisfying the criterion for the convergence,

$$\left| c(r)^{new} - c(r)^j \right|_{max} \leq 10^{-5} \tag{10}$$

g. If $c(r)^{new}$ satisfies the criterion then it is the expected solution. However, if it does not satisfy then a new input for the $(j+1)^{th}$ iteration has to be generated before going to step (b).

The speed of convergence of the iterative procedure is very sensitive to the new input for the successive iteration. One method to generate the $(j+1)^{th}$ input is by mixing the jth and the 'new' values as

$$c(r)^{j+1} = xc(r)^j + (1-x)c(r)^n \tag{11}$$

Where x is a number having a value between 0 and 1. The function of x is to avoid large differences between successive iterates and therefore to prevent divergence.

In order to check the accuracy of our method of calculation, we have compared in Figure 1 the radial distribution function obtained by solving the RY-integral equation theory to the Brownian dynamic simulation[8] for a system of colloids interacting via a potential varying as $1/r^3$. We see here that for this system the results obtained by RY-theory is in excellent agreement with the simulation. It provides us the confidence regarding the accuracy of the procedure of the solution.

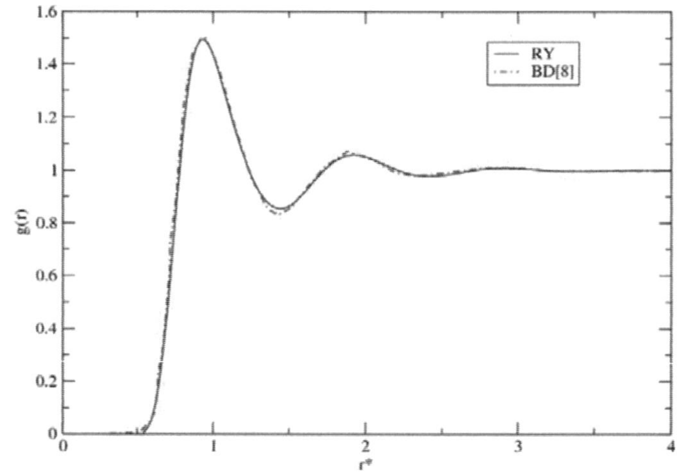

Figure 1. Plot representing comparison of g(r) calculated by Roger-Young's closure and Brownian dynamic simulation[8] at $\Gamma = 1.44$ in 2D system for the potential $\frac{\Gamma}{r^3}$.

IV. Results and Discussion

2D Case

Γ	$\frac{\beta P^v}{\rho}$	$\frac{\beta P^c}{\rho}$	$\frac{P^c}{P^v}$
10	66.21	69.84	1.054
15	96.68	98.77	1.021
20	129.01	128.67	0.997

Table 1. Comparison of pressure from virial and compressibility route for 2-dimension.

In Table 1, we have tabulated the values of pressure of a system in two-dimensional plane, calculated by using virial and compressibility roots for various values of strength parameter. We found that for one particular value of alpha the values of pressure calculated through both the routes are almost same. Hence we conclude that for this value of alpha, the requirement of self-consistency is satisfied.

In Figure 2 , we have plotted the radial distribution curve for the case of 2D. We observe again that the peak height increases with the increase in strength of repulsive interaction. One interesting feature to be mentioned here is the relative heights if peaks and the structures in 3D and 2D radial distribution functions. We found here that for the same strength of repulsive interaction, the radial distribution function for 2D case is more intense and of longer range than that of 3D case

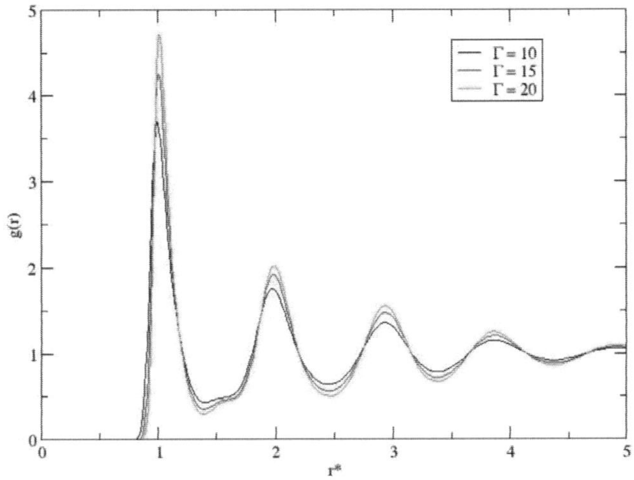

Figure 2. Plot of radial distribution function g(r) at Γ =10, 15 and 20 for the case of 2D

A. 3D case

Γ	$\frac{\beta P^v}{\rho}$	$\frac{\beta P^c}{\rho}$	$\frac{P^c}{P^v}$
10	102.082	103.57	1.014
15	152.99	151.13	0.987
20	200.28	198.04	0.988

Table2 . Comparison of pressure from virial and compressibility route for 3-dimensio

In Table 1, we have listed the virial and compressibility pressures for the system in bulk. We observe here that the values of pressure in 3D are higher than that for 2D case. Again for a particular value of mixing parameter the virial and compressibility pressures are very close to each other and hence this value of mixing parameter may be considered for the solution of OZ.

Using above value of alpha in Roger-Young closure we solved the OZ equation. In Figure 1 we have plotted the radial distribution function corresponding to Gamma=10, 15, and 20. We observe that the peak height of g(r) increases with the increase in Gamma. This indicates enhanced local ordering with the increase in the strength of repulsive interaction.

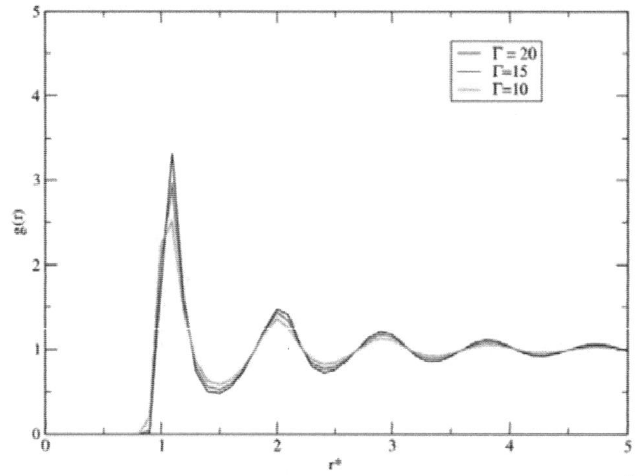

Figure 3. Plot of radial distribution function g(r) at Γ =10, 15 and 20 for the case of 3D

V. Conclusions

In this work we have determined , the pair structures of a system of particles interacting confined in 3D/2D space and interacting via a potential varying as 5th power of the interparticle separation. We have solved Roger-Young integral equation theory by determining the mixing parameter which satisfy the consistency of virial and compressibility pressure. We have presented the radial distribution curves of 2D and 3D case and have shown that the radial distribution curves of 2D case is not only nor intense but exhibit longer range. In future we intend to use the pair structure in density functional theory to determine the topology of the phases exhibited by quadrupole system.

Acknowledgment

B.K.M and A.K would like to thanks Indian School of Mines, Dhanbad for providing financial supports through research fellowship.

References

[1] Hansen J.P and McDonald I.R. 1986 *Theory of simple liquids*, Academic Press.London.2nd edition.

[2] Caccamo .C 1996 Phys. Rep. 274,1 .

[3]Born M and Green H S, Proc R Soc.London,Ser. A 1946 188,10.

[4] Percus J K and Yevick G J 1958 Phys.Rev.110, 1

[5]Rosenfeld Y and Ashcroft N Y, 1979 *Theory of simple classical fluids: Universality in the short-range structure* 20, Phys.Rev.A, 3.

[6]Hus M, Zalar M , Urbic T 2013 ,138 J. Chem. Phys. 224508,

[7]Singh R C, Ram J and Singh Y 1996 *Thermodynamically self-consistent Integral-equation theory for pair-correlation functions of a molecular Fluid* 54 Phys. Rev. E 1.

[8]Hoffmann N, Likos C N and Lowen H 2006*Microphase structuring in two –dimensional magnetic colloid mixtures* 18 J.Phys.:Condense. Matter

Role of magnesium on the biomimetic deposition of calcium phosphate

Bimal K. Sarma* and Bikash Sarma

Nanophysics Laboratory, Department of Physics, Gauhati University, Guwahati-781014, Assam, India

*E-mail: bimal@gauhati.ac.in, sarmabimal@gmail.com

Abstract Biomimetic depositions of calcium phosphate (CaP) are carried out using simulated body fluid (SBF), calcifying solution and newly developed magnesium containing calcifying solution. Calcium phosphate has a rich phase diagram and is well known for its excellent biocompatibility and bioactivity. The most common phase is hydroxyapatite (HAp), an integral component of human bone and tooth, widely used in orthopedic and dental applications. In addition, calcium phosphate nanoparticles show promise for the targeted drug delivery. The doping of calcium phosphate by magnesium, zinc, strontium etc. can change the protein uptake by CaP nanocrystals. This work describes the role of magnesium on the nucleation and growth of CaP on Ti and its oxide substrates. X-ray diffraction studies confirm formation of HAp nanocrystals which closely resemble the structure of bone apatite when grown using SBF and calcifying solution. It has been observed that magnesium plays crucial role in the nucleation and growth of calcium phosphate. A low magnesium level enhances the crystallinity of HAp while higher magnesium content leads to the formation of amorphous calcium phosphate (ACP) phase. Interestingly, the deposition of ACP phase is rapid when magnesium ion concentration in the solution is 40% of calcium plus magnesium ions concentration. Moreover, high magnesium content alters the morphology of CaP films.

1. Introduction

Calcium phosphate (CaP) is a bioceramic and the synthetic forms have the potential for biomedical applications [1–3]. Magnesium is a vital component of bone mineralization. Bone mineralization is a complex phenomenon and so extremely challenging to mimic the process completely [4, 5]. Initially, authors of this work tried to use simulated body fluid (SBF) [6], but found the deposition to be very slow (requires at least 4 weeks). Since the work reported here compares the growth of calcium phosphate, a slightly concentrated SBF (1.5SBF), a calcifying (Ca/P) solution [7], a modified calcifying (m-Ca/P) solution and magnesium containing m-Ca/P solution are used. Even higher concentration of SBF, i.e. 5SBF is a natural choice for faster deposition of CaP. However, the crystallinity of apatite is reduced when grown using 5SBF [8]. Another important aspect is unlike CaP grown using 1.5SBF, Ca/P solution, m-Ca/P solution where deposition takes place uniformly, apatite grown in 5SBF is less uniform and show evidence of poor adhesion with the substrate [8, 9].

In this work, biomimetic deposition of calcium phosphate with a controlled incorporation of magnesium from a low to high level of Mg^{2+} ion concentration in a modified calcifying solution is presented. Understanding the effect of magnesium on the nucleation and growth of calcium phosphate on implant materials is the central theme of this work.

2. Experimental methods

2.1. Preparation of solution

The ion concentrations of human blood plasma and different solutions used in this work are presented in table 1.

Table 1. Ion concentrations (mM) of human blood plasma and different solutions for the deposition of CaP films

Ion	Human blood plasma	SBF	1.5SBF	Calcifying solution Ca/P	Modified Calcifying solution m-Ca/P	m-Ca/P with Mg content
Na^+	142.0	142.0	213.0	25.5	76.5	76.5
K^+	5.0	5.0	7.5	—	—	—
Mg^{2+}	1.5	1.5	2.25	—	—	5–0.625
Ca^{2+}	2.5	2.5	3.75	2.5	12.5	7.5–11.875
Cl^-	103.0	147.8	221.7	5.0	25.0	25.0
HCO_3^-	27.0	4.2	6.3	18.0	54.0	54.0
HPO_4^{2-}	1.0	1.0	1.5	—	—	—
PO_4^{3-}	—	—	—	2.5	7.5	7.5
SO_4^{2-}	0.5	0.5	0.75	—	—	—

The reagents NaCl, $NaHCO_3$, KCl, $K_2HPO_4.3H_2O$, $MgCl_2.6H_2O$, $CaCl_2$, Na_2SO_4 and $Na_3PO_4.12H_2O$, as received from Merck and Sigma Aldrich are used to prepare the solutions in de-ionized water. The solutions are prepared at 37.0±0.5 °C using a heating magnetic stirrer (Arex Digital Pro, Velp Scientifica). The pH of the solutions at each stage is measured by using a pH meter (Cyberscan Ph2100, Eutech Instruments). Finally, pH is adjusted within the range 7.3–7.4 by adding required amount of Tris or Hepes and HCl and maintaining temperature of the solution at 37 °C. The detailed procedure is described elsewhere [8, 9]. The films are grown at a temperature of 37 °C.

2.2. Recording variation of pH with time

The pH variation of different solutions with substrate immersion time is recorded using the Cyberscan Ph2100 pH meter interfaced with the CyberComm Pro Data Acquisition Software.

2.3. Characterization of prepared samples

The surface morphology and elemental analyses of the samples are examined using field emission scanning electron microscope (FE-SEM, ΣIGMA, Zeiss) equipped with energy dispersive x-ray (EDX) spectrometer. XRD patterns are recorded in the powder diffractometer (D8 Advance, Bruker AXS). The Cu K_α line 1.5406 Å is used at a low grazing angle of 1°. The intrinsic broadening due to the diffractometer is corrected by recording the XRD pattern of the standard corundum sample under identical conditions. XRD patterns of the samples are acquired in the 2θ range 20–55° and step size of 0.02° and integration time of 1 s/step.

3. Results and discussion

3.1 pH variation

The pH variation (figure not shown here) of different solutions with Ti substrate immersion time is investigated. The starting pH values are 7.38 and burgeon during first 3 hours. With time, pH gradually increases. The increase in pH of a solution is primarily due to the release of CO_2 gas as [8]

$$NaHCO_3 \rightarrow Na^+ + HCO_3^- \quad (1)$$

$$HCO_3^- \rightarrow CO_2 + OH^- \tag{2}$$

With increase in pH, respective solutions achieve supersaturation which in turn triggers deposition of calcium phosphate. However, the increase in pH of solutions occurs at different pace for different solutions. Since these changes happen at the early stages of deposition of CaP so provides an initial impression that the precipitations of CaP phase might not be the same in all cases.

3.2 Microstructure

Figure 1. High resolution FE-SEM images of CaP films grown on Ti substrate using (a) 1.5 SBF, (b) 5.0 SBF, (c) calcifying solution, (d) modified calcifying solution, (e) modified calcifying solution with moderate Mg content, (f) modified calcifying solution with high Mg content

The surfaces of CaP structures grown on Ti using various solutions are composed of uniformly distributed micron sized spherical particles. Higher resolution FE-SEM images (Figure 1) show that the surfaces of CaP consist of nano-sized plates arranged to form flowery structures. These plate shaped tiny particles are closely connected to each other and are characteristics of bone apatite [6–8]. Interestingly, such morphology prevails in CaP structure at varying Mg^{2+} concentrations in the m-Ca/P solution from low to moderate level. However, higher magnesium content (40% of calcium plus magnesium ions concentration) alters the morphology drastically to globular pattern signifying that the magnesium content in solution plays a crucial role in the precipitation of CaP structure. EDX results are summarized in table 2. Ca/P atomic percentage ratio deviates from that of the stoichiometric hydroxyapatite (HAp) which is 1.67. However, the calcium deficiency cannot be solely attributed to the magnesium substitution at the Ca site in the HAp lattice. Calcium deficient carbonate apatite is quite common in biomimetic deposition of CaP.

3.3 X-ray diffraction and analysis

Figure 2A shows XRD patterns of biomimetic CaP films grown on Ti substrate using 5 SBF, 1.5SBF, Ca/P, m-Ca/P, and m-Ca/P solutions with Mg source. X-ray line profile fitting for XRD pattern of CaP film with 7.2% Mg to Ca atoms concentration (corresponding to 10% of magnesium to calcium plus magnesium ions concentration in m-Ca/P solution, i.e. m-Ca/P solution with moderate Mg content) is shown in figure 2B. XRD patterns reveal formation of HAp phase in all cases except for CaP films grown using m-Ca/P solution with high Mg content (corresponding to 40% of magnesium to calcium plus magnesium ions concentration in m-Ca/P solution, i.e. 29.6% Mg to Ca atoms concentration) which is amorphous (table 2).

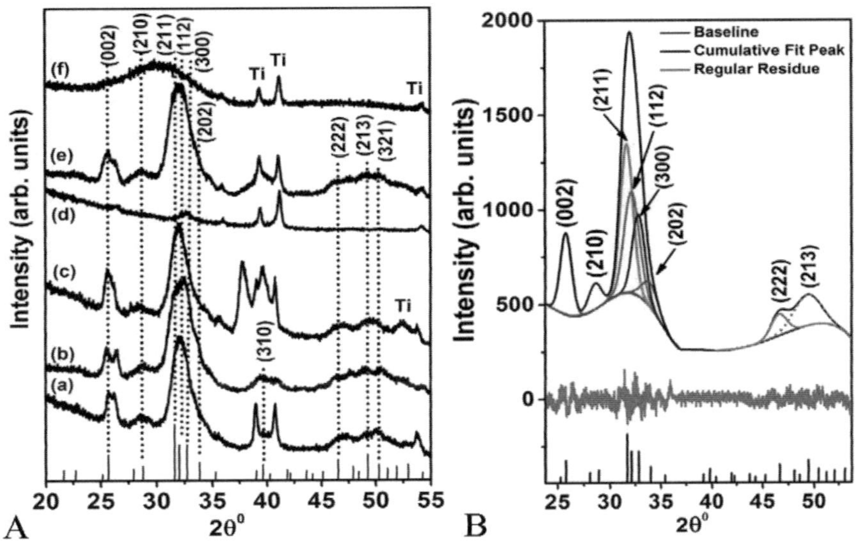

Figure 2. A. XRD patterns of CaP films grown on Ti substrate using (a) 1.5 SBF, (b) 5.0 SBF, (c) Ca/P solution, (d) m-Ca/P solution, (e) m-Ca/P solution with moderate Mg content, (f) m-Ca/P solution with high Mg content, B. Representative X-ray line profile fitting for HAp with moderate Mg content.

Table 2. Mg content, phase, and crystallite size of biomimetic CaP films

Solution	% Mg/Ca	% Mg/ (Ca+Mg)	Phase	Ca/P	(Ca+Mg)/P	*t* (nm)
Ca/P	0	0	Carbonate HAp	1.35	1.35	9
m-Ca/P	0	0	Carbonate HAp	1.49	1.49	3
m-Ca/P with Mg content	4.8	4.6	Carbonate HAp	1.34	1.40	6
	7.2	6.7	Carbonate HAp	1.43	1.54	6
	7.7	7.1	Carbonate HAp	1.55	1.67	8
	29.6	22.9	ACP	1.07	1.38	—

HAp is nanocrystalline with apparent domain size in the range 3–9 nm. The result is consistent with TEM observation (data not presented here). The appearance of the most intense peak at ~32° might be ascribed to the collective contributions from (211), (112), (300), and (202) diffraction planes which is obvious from the X-ray line profile fitting as shown in figure 2B. Raman spectra (figure not shown here) further suggest that nanocrystalline HAp has B-type substitution with carbonate ions replacing some of the phosphate ions in the HAp. At the same time, with Mg source in the solution, Mg substitutes for Ca in the HAp lattice. EDX analyses (table 2) suggest 4.8–7.7% substitution of Mg to Ca in the nanosized carbonate apatite. With greater Mg content (29.6% Mg to Ca atoms) the amorphous calcium phosphate (ACP) phase is formed. In order to understand the effect of Mg substitution, the unit cell parameters are determined from X-ray line profile fittings for various Mg concentrations. The variation is shown in figure 3.

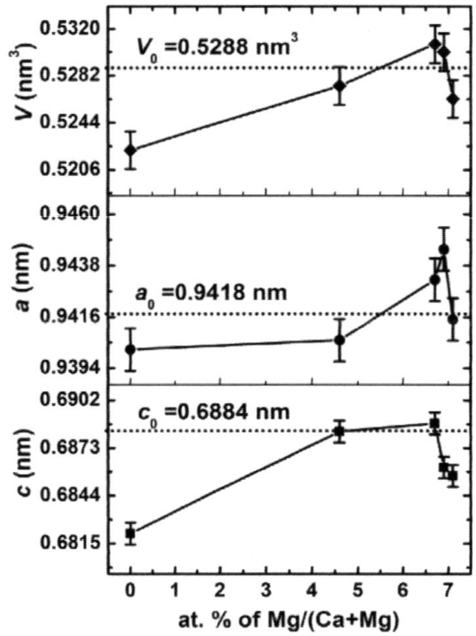

Figure 3. Variations of lattice parameters *c*, *a*, and unit cell volume *V* of CaP as a function of Mg impurity concentration. The dashed lines represent standard values for a, c, and V taken from ICDD file 9-432.

It is observed that the unit cell initially experiences extension along *a*–and *c*–directions of the hexagonal HAp with increasing Mg concentration. However, beyond 6.7% of Mg concentration, the unit cell undergoes substantial contraction in the *c*-direction. However, the lattice contraction along *a*–direction is not so significant even at 7.1% of Mg concentration. The unit cell volume follows a similar trend. Interestingly, such fluctuations in *a*, *c*, and *V* are observed at different Mg concentrations even with equal apparent domain size (6 nm). The evident lattice distortion is better decided by the impurity concentration. Even if the domain sizes are equal, marked variations in *a*, *c*, and *V* are noticed.

4. Conclusions

Biomimetic approach to CaP nanocrystals reveal that a modified calcifying solution may be suitably used for Mg doping experiments. The growth of HAp nanocrystals on Ti substrate from a supersaturated calcifying solution at physiological temperature and pH of human blood plasma has been demonstrated through XRD, Raman, and electron microscopy studies. The apparent domain size lies in the range 5–8 nm with low to moderate Mg concentration. Interestingly, such extent of magnesium substitution does not inhibit the HAp crystal growth. A controlled Mg incorporation could induce lattice distortion in the carbonate apatite. However, higher Mg level is found to be responsible for the precipitation of ACP films. Moreover, the precipitation of ACP is much faster than HAp growth suggesting that the role of Mg is more varied here and is not just limited to inducing the lattice distortion in the HAp lattice. It is well known that Mg is important for human metabolism and magnesium deficiency in bone can lead to orthopedic problems. Thus the presented method of controlled incorporation of Mg in biocompatible CaP might be a potential way to produce next generation biomaterials for biomedical applications.

Acknowledgments
BKS acknowledges the financial support from the University Grants Commission (UGC), New Delhi, India under the Grant No. F. 30–20/2014 (BSR). The authors greatly appreciate the technical support of Dr. A R Pal and his team of the Institute of Advanced Study in Science and Technology, Guwahati for the technical support.

References
[1] Williams D F 2008 *Biomaterials* **29** 2941–53

[2] Altinoglu E and Adair J H 2009 *Future Oncol.* **5** 279–81

[3] Webster T J and Ejiofor J U 2004 *Biomaterials* **25** 4731–9

[4] Taton T A, 2001 *Nature* **412** 491–2

[5] Nudelman F, Pieterse K, George A, Bomans P H H, Friedrich H, Brylka L J, Hilbers P A J, G. With, and Sommerdijk N A J M 2010 *Nat. Mat.* **9** 1004–9

[6] Kokubo T and Takadama H 2006 *Biomaterials* **27** 2907–15

[7] Bigi A, Boanini E, Bracci B, Facchini A, Panzavolta S, Segatti F and Sturba L 2005 *Biomaterials* **26** 4085–9

[8] Sarma B K, Das A, Barman P, Pal A R 2016 *J. Phys. D: Appl. Phys.* **49** 145304

[9] Sarma B K, Barman P, Sarma B, Das A, Pal A R 2016 *Mater. Lett.* **185** 387

Photoluminescence Behaviour of Sm^{3+} Ions in presence of Ag Nanoparticles in Methanol

N Dehingia[1], P Gogoi[2], A Boruah[1], D Kakoti[1], N Rajkonwar[1] and *P Dutta[1]
[1]Laser and Spectroscopy Laboratory, Department of Physics, Dibrugarh University Dibrugarh, India,
[2]Dibrugarh University Institute of Engineering and Technology
*pankaj@dibru.ac.in

Abstract. In the present work, capped Ag NPs prepared by reduction of Ag $(NO_3)_3$ by Dimethyl Formamide is doped with Sm^{3+} in methanol and its photoluminescence behavior is studied. Significant modifications of the Sm^{3+} ions' emission as well as quantum yield, were observed with the concentration of Ag NPs. Local field enhancement induced by neutral Ag NPs were found to be responsible for enhancement in efficiency of the Sm^{3+} ions.

1. Introduction

Photoluminescence (PL) of rare earth (RE) ions is one of the principal technologies that support laser amplification, optical communication, sensing, display systems etc. However, the small absorption cross section of RE ions has spawned numerous attempts to enhance the PL efficiency of these ions [1]. Coupling of RE ions with metal nanoparticles (NPs) have been emerged as a novel technique to enhance the PL intensity of RE ions [2].

Metal NPs possess fascinating optical and electronic etc. properties compared to their bulk counterparts. Electromagnetic excitations induce noble metal NPs to exhibit the Surface Plasmon Resonance (SPR), which is the collective oscillation of the free conduction band electrons in the visible region. SPR generates enhanced localized electric field around the metal NPs. RE ions, when present within the close proximity of such NPs, experience a drastic increase in their excitation rates by the induced electric field and thereby their spectroscopic properties get modified [3]. The shape, size, number density of metal NPs and the coupling distance between the NP and RE ion also have a crucial role on the PL enhancement from the RE ions [4].

In the present work, the effect of Silver (Ag) NPs on the PL behavior of Samarium (Sm) ions in methanol has been investigated.

2. Experimental

Stock solutions of Silver Nitrate ($AgNO_3$) (Acros Organics, 99.85%) and Polyvinylpyrrolidone (PVP) (SRL, research grade) were first prepared separately by dissolving $AgNO_3$ in Dimethyl Formamide (DMF) (Acros Organics, 99%) and PVP in distilled water respectively at room temperature. Both the solutions were then mixed and kept for four hours. Samarium Nitrate hexahydrate (Aldrich, 99.99%) was mixed with the solution containing the other precursors methanol, distilled water and nitric acid in the volumetric ratio of 2: 3: 0.4. This solution was mixed with the stock solution of $AgNO_3$ and magnetically stirred for 30 minutes. Altogether four sets of samples were prepared for fixed concentration of Samarium Nitrate hexahydrate and varied concentrations of $AgNO_3$. Composition of the samples used for the study is shown in Table 1.

Table 1. Composition of the prepared samples.

Sample Name	$SmNO_3.6H_2O$ (wt%)	$AgNO_3$ (wt%)
S	2.24	-
SA1	2.24	0.003
SA2	2.24	0.03
SA3	2.24	0.3

The UV-Vis-NIR absorption spectra of the prepared samples were recorded by using Scinco PD Array (S-3100) UV-Vis spectrophotometer in the wavelength range of 200-1100nm. PL and PL excitation spectra were recorded by Fluoromax 4P spectrofluorometer equipped with 150W CW xenon lamp as excitation source and photomultiplier tube detector. The samples were excited at 402nm keeping the excitation and emission slits both at 3nm. All the measurements were done at room temperature.

3. Results and discussions

Fig.1 shows the UV-Vis-NIR absorption spectra for $AgNO_3$ doped with Sm ions in methanol. Absorption spectrum for $AgNO_3$ singly doped sample (with different scale) is also incorporated in the figure for comparison. To show the resolved Sm^{3+} peaks, the enlarged view of the absorption spectra for samples S, SA1 and SA2 in 350-550 nm range are shown in Fig. 2.

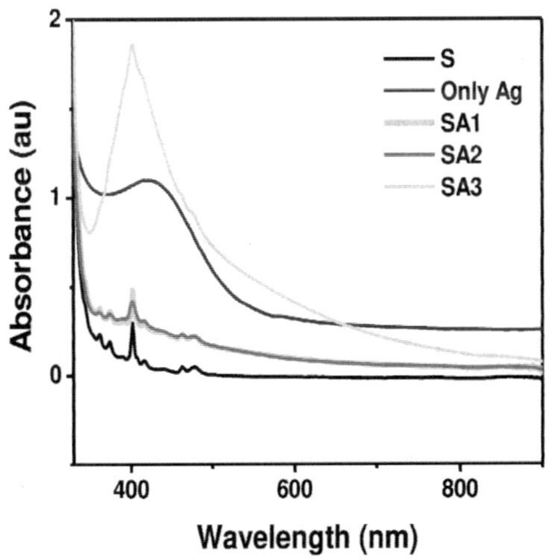

Figure 1. Absorption spectra for samples S, SA1, SA2, SA3 and singly doped $AgNO_3$.

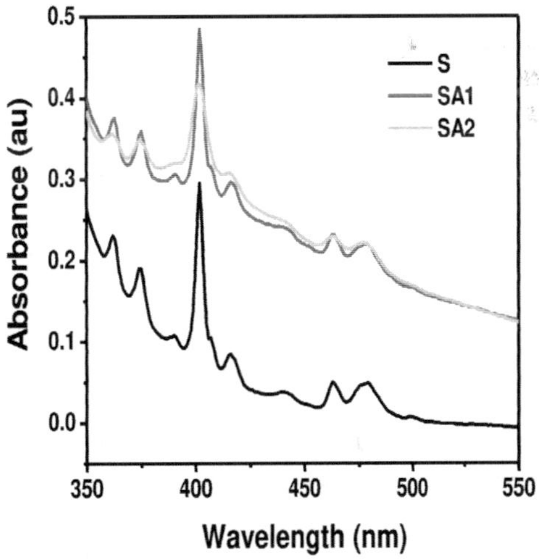

Figure 2. Enlarged absorption spectra for samples S, SA1& SA2.

The spectra for S, SA1 and SA2 show seven absorption peaks located around 362, 374, 388, 402, 416, 463 and 479nm respectively arising due to the f-f transitions from the ground state $^6H_{5/2}$ of Sm^{3+} [5]. It is observed that the absorption intensity as well as FWHM for Sm^{3+} peaks first increases with the addition of $AgNO_3$. The enhancement was observed till the $AgNO_3$concentration of 0.03 wt%. With further increase in the concentration of $AgNO_3$ to 0.3 wt% (SA3), the SPR band of Ag at 416 nm broadened, became asymmetric and the tail extended upto 800nm [6].

Fig. 3 illustrates the PL spectra of the samples under the excitation wavelength of 402 nm. The spectra show two intense PL bands around 596nm and 642nm in the visible region together with a weak band at 560nm and are assigned as $^4G_{5/2} \rightarrow {}^6H_{5/2}$ (green), $^4G_{5/2} \rightarrow {}^6H_{7/2}$ (orange), $^4G_{5/2} \rightarrow {}^6H_{9/2}$ (red) transitions respectively [5].

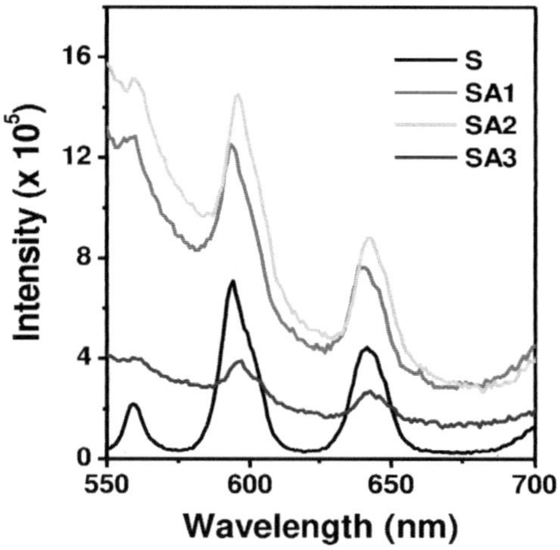

Figure. 3: PL spectra for samples S, SA1, SA2 & SA3 under 402nm excitation.

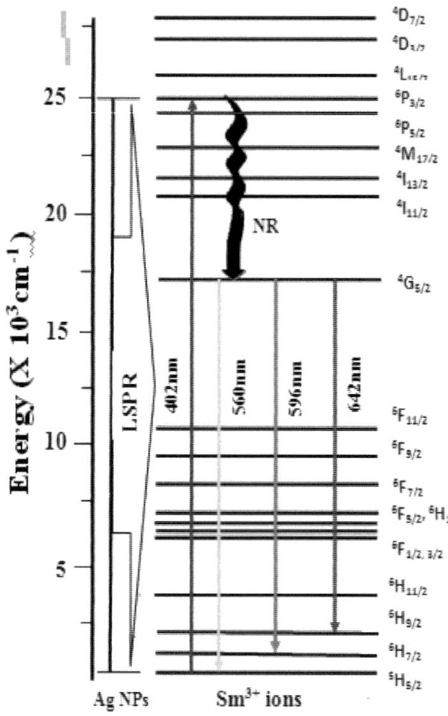

Figure 4. Schematic energy level diagram of Sm^{3+}: Ag NPs under 402nm excitation.

It is seen from Fig. 3 that the PL intensity first increases with the increase in concentration of $AgNO_3$ (till 0.03 wt%). Further increase in concentration to 0.3 wt% results decrease in the emission intensity. The PL enhancement of the Sm^{3+} ions in the presence of Ag NPs may be attributed to the Local field enhancement (LFE) around the Ag NPs induced by the SPR. Consequently, the electric field around the RE ions, lying within the proximity of metal NPs enhances. This in turn increases the rate of excitation of the RE ions and hence the PL intensity of Sm^{3+} ions is enhanced. Moreover, with the increase in concentration of $AgNO_3$, the electric fields of the NPs overlie and produce hot spot regions that lead to stronger surface enhanced fluorescence. However, beyond certain concentrations of Ag NPs (0.3 wt% for present case), an undesired back transfer of energy from Sm^{3+} to the nanoparticles might occur causing the decrease in emission intensities. Also, since a fraction of Ag NPs may still remain as Ag^+ ions, dimers and multimers in the system, energy absorbed by Sm^{3+} may get transferred to the Ag^+ ions resulting in the quenching of PL [7].

Fig. 4 shows the schematic energy level diagram for Sm^{3+} ions in the vicinity of Ag NPs. Under 402nm excitations, the 4f electrons of Sm^{3+} ions at the ground state are excited to the $^6P_{3/2}$ state by single step ground state absorption (GSA). The electrons then decay non-radiatively and populate the lower $^4G_{5/2}$ level. Radiative transitions from $^4G_{5/2}$ level to $^6H_{5/2}$, $^6H_{7/2}$ and $^6H_{9/2}$ produce respectively the observed green emissions at 560nm, orange emission at 596nm and red emission at 642 nm.

Table 2. Comparison of PL intensities, quantum yields and FWHMs for Sample S, SA1, A2 & SA3.

PL maxima	PL Intensity			
	S	SA1	SA2	SA3
560	1.19 x10^5	1.28 x10^6	1.51 x10^6	3.98 x10^5
596	7.1 x10^5	1.25 x10^6	1.45 x10^6	3.89 x10^5
642	1.77 x10^5	7.65 x10^6	8.81 x10^5	2.71 x10^5
		Relative Yield		
560	2.04 x10^6	1.08x10^7	1.44x10^7	4.26x10^6
596	1.04 x10^7	2.1x10^7	2.52x10^7	6.07x10^6
642	7.12x10^6	1.38x10^7	1.5x10^7	4.8x10^6
		FWHM (nm)		
560	8	9	10	11
596	14	21	22	19
642	14	23	22	22

4. Conclusions

In the present work, absorption and PL behavior of Sm^{3+} ions in presence of Ag NPs in methanol medium were studied. An enhancement of the PL intensities by two times and yields by 2.4 times with the addition of 0.03 wt% of $AgNO_3$ were observed which shows a decrement when the concentration was further enhanced to 0.3 wt%. The study reveals that localized SPR is responsible for the enhancement of the absorption and PL efficiencies of the Sm^{3+} ions whereas the decrement in efficiency is attributed to the back energy transfer from Sm^{3+} ions to the Ag NPs in the studied medium.

References

[1] Dousti M R, Amjad R J and Mahraz Z A S 2015 *J. Mol. Stuctr.* **1079** 347.
[2] Rivera V A G, Ledemi Y, Osorio S P A, Manzani D, Messaddeq Y, Nunes L A O and Marega Jr E 2011 *J. Non-Cryt. Solids* **358** 399.
[3] Pillonnet A, Berthelot A, Pereira A, Derom S, Colas des Francs G and Jurdyc A M 2012 *Appl.Phys. Lett.* **100**, 1531151.
[4] Yanagida S, Nakagawa T, Kishi T and Yasumori A 2015 *J. Sol-gel Sci. Technol.* **74** 227.
[5] Babu B H and Ravi Nath Kanth Kumar V V 2012 *J. Appl. Phys.* **112** 093516.
[6] Som T and Karmakar B, 2009 *J. Appl. Phys.* **105** 0131021.
[7] Fares H, Elhouichet H, Gelloz B and Ferid M 2015 *J. Appl. Phys.* **117** 193102.